Mathematics of Chance

JIŘÍ ANDĚL
Charles University
Czech Republic

A Wiley-Interscience Publication
JOHN WILEY & SONS, INC.
New York • Chichester • Weinheim • Brisbane • Singapore • Toronto

This text is printed on acid-free paper. ♾

For ordering and customer service, call 1-800-CALL WILEY.

Library of Congress Cataloging in Publication Data is available.

ISBN 0-471-41089-6

Printed in the United States of America

10 9 8 7 6 5 4 3 2 1

Mathematics of Chance

WILEY SERIES IN PROBABILITY AND STATISTICS
TEXTS, REFERENCES, AND POCKETBOOKS SECTION

Established by WALTER A. SHEWHART and SAMUEL S. WILKS

A complete list of the titles in this series appears at the end of this volume.

Contents

List of Figures

List of Tables

Preface

I have compiled some interesting and popular elementary problems concerning probability theory and mathematical statistics for a long time, and I use them in this book as examples in fields of mathematics discussed here. In 1998 I published a Czech preprint of this book. I sent copies to my colleagues, who called my attention to misprints and errors in the material and suggested many modifications and improvements. In the English translation I omit a few topics that may be of interest to Czech readers only, such as prices of marked research in the Czech Republic and Czech Parliament elections results. I also omit the final original chapter on queuing theory. However, I include some new problems that I found in the meantime.

Most of the material presented here is accessible to first- and second-year undergraduate students who are acquainted with the basic elements of calculus. There are only a few places in the book where some knowledge of mathematical statistics is needed. I do not present statistical methods for data analysis. Such procedures, with interesting examples and applications, can be found in other publications. Rather, I use the basic elements of probabilistic and statistical reasoning in my explanations. The end of a proof is denoted by the symbol □, as is usual in mathematical texts. I think that this will improve the overall organization of the book.

J. ANDĚL

Prague, Czech Republic

Acknowledgments

I would like to express my thanks to many colleagues who read the Czech preprint and some of its previous versions and suggested many improvements and offered encouragement. I am especially grateful to Václav Dupač, Jan Hurt, Josef Machek, Ivan Netuka, and Břetislav Novák, who devoted much time and effort to develop the material presented here. I also owe gratitude to Jaromír Antoch for his help with preparing LaTeX version of this book.

I am grateful to the editorial and production staff at John Wiley & Sons, Inc. I am particularly indebted to Heather Haselkorn (Editorial Program Coordinator), Steve Quigley (Acquisitions Editor), and Melissa Yanuzzi (Production Editor).

Financial support from grants GAČR 201/97/1176, GAČR 201/00/0770, CEZ:J13 /98:113200008, and MSM113200008 in the preparation and compilation of this material and for publication of the Czech and English preprints is also gratefully acknowledged.

J. A.

Introduction

The usefulness of probability theory is recognized throughout the world as well as in the Czech Republic. Decisions made with uncertainty must be based on statistical data. I introduce an example from our Faculty of Mathematics and Physics (FMP) at Charles University in Prague. Data from the past few years (at the time of writing) indicate that only 60% of students accepted for the first academic year actually enroll, as some students apply for faculties besides FMP. If such students successfully pass entrance examinations at other faculties, they generally enroll there and are not longer interested in study at FMP. If the enrollment capacity of FMP is 120 students in first-year mathematics, then elementary computation indicates that the dean should accept about 200 applicants for mathematics.

Thus it does not seem very difficult to use compilled statistical data. However, this is not true in most cases. Even such elementary characteristics as the average has some properties that can surprise a nonstatistician (and sometimes also a statistician). We introduce two short cases.

In case 1 (see *Teaching Statistics* **13**, 1991, p. 23); we should realize that most people have more than the average number of feet. In the second case (see *Teaching Statistics* **14**, 1992, p. 29); it is well known that English and Scottish citizens mutually joke about each other. In an Edinburgh pub this comment was overheard: "A Scottish person who moves from Scotland to England raises the levels of intelligence in both places." One can ask following questions.

1. Is this possible?

2. What might you hear in a Newcastle pub?

The answer to the first question is affirmative. Imagine, for simplicity, that there are only two English and two Scottish people, A_1 and A_2 and S_1 a S_2, respectively. People in Edinburgh probably think that intelligence level is distributed like this:

	A_1	A_2	S_1	S_2
IQ	100	120	140	160

The average English and Scottish intelligence quotients (IQs; on the Stanford Binet scale) are 110 and 150, respectively. If S_1 moves to England (and the Scottish believe that only a person with lower IQ would do this) the average IQ in England will be 120. Only S_2 remains in Scotland, so the average IQ there is 160. Thus the level of intelligence raises in both places. It is easy to modify these data to arbitrary large numbers of English and Scottish people.

And what might you hear in a Newcastle pub? We can imagine that people living there claim that only a brighter Scot would think of moving, and so intelligence decreases in both countries. Of course, people in Newcastle believe that the IQ distribution is as follows:

	S_1	S_2	A_1	A_2
IQ	100	120	140	160

Thus English believe that average Scottish and English IQ levels are 110 and 150, respectively. If S_2 moves to England, then the average Scottish IQ is 100 and the English IQ, 140. Thus the average IQ decreases in both places.

If one deals with characteristics other than averages, the situation can be even more complicated. Assume that the male/female ratio is studied in two districts, say, in districts A and B. The inhabitants are divided into two groups — younger and older, as follows:

	District A			District B		
Category	Women	Men	Total	Women	Men	Total
Younger	5	6	11	3	4	7
Older	6	3	9	9	5	14
Total	11	9	20	12	9	21

A reader who thinks that the number of inhabitants in the districts is too low can multiply all numbers by 10^5. The results of the following analysis remain the same.

The ratio of younger women is higher in district A since

$$\frac{5}{11} = 45.5\% > \frac{3}{7} = 42.9\%.$$

The same is true for older women, since

$$\frac{6}{9} = 66.7\% > \frac{9}{14} = 64.3\%.$$

If we consider totals, then we can see that the ratio of all women in district A is lower than in district B, since

$$\frac{11}{20} = 55.0\% < \frac{12}{21} = 57.1\%.$$

This is an illustration of the *Simpson paradox*, which is expressed mathematically as follows. Let a_i, b_i, c_i, and d_i, $i = 1, 2$, be positive numbers. Then

$$\left\{\frac{a_1}{b_1} < \frac{c_1}{d_1}, \frac{a_2}{b_2} < \frac{c_2}{d_2}\right\} \not\Rightarrow \left\{\frac{a_1 + a_2}{b_1 + b_2} < \frac{c_1 + c_2}{d_1 + d_2}\right\}.$$

This result is perhaps not very striking. But numerical illustrations (such as the one mentioned above) are usually quite surprising.

We introduce another example of faulty interpretation of statistical data. Dave Allen formulated it as follows. "Statistics show that 10% of all road accidents are caused by drunken drivers, which means that the other 90% are caused by sober drivers, ... so why can't they get off the road and leave us drunks to it?" (*Teaching Statistics* **6**, 1984, p. 52).

1

Probability

1.1 INTRODUCTION

Most events that we deal with have a random character. For example, one hour may or may not suffice for a visit to a physician, or if we turn on a light, we cannot be sure that the bulb will work all evening. Such examples abound indefinitly.

The randomness of events has been known for a long time. Dice and playing cards have a long history. Remember the statement by Gaius Julius Caesar: "Alea iacta est" ("dice are cast").

One of the old problems, called the *problem of division*, concerns calculation of probability. Two players, say, A and B, play a series of games. Assume that there are no draws (or that draws are not taken into account). The play is *fair*, which means that A wins a game with the same probability as B, that is, with probability $\frac{1}{2}$. The games are independent. The first player who wins six games is the winner and gets a prize. The rules are similar to those that sometimes apply for world championship in chess. The play is interrupted at the moment when A won five games and B won three games. Since there is no chance to continue, our problem is how to divide the prize fairly. This problem was solved already by Fra Luca Paccioli [1445(?)–1514(?); sometimes the year of his death is given as 1509] in the book *Summa de arithmetica, geometria, proportioni et proportionalita*, which was published in Venice in 1494. Fra Luca Paccioli was a friend of Leonardo da Vinci (1472–1519). Székely (1986) writes that an old Italian manuscript from 1380 was found where the same problem was introduced. It now seems to be of Arabian origin. Paccioli did not realize that the problem is probabilistic in nature. He used some formal calculations and erroneously concluded that the fair division of the prize is 2:1. An erroneous solution was also published by Niccolo Tartaglia [1499(?)–1557] in the book *General trattato di numeri et misura*, which appeared in 1556. Remember that this author was able to find in one night a formula for the solution of the cubic equation.[1] Only Blaise Pascal (1623–1662) and Pierre de Fermat (1601–1665) independently and in different ways derived the correct result in 1654. They came to the surprising conclusion that the fair division of prize is in the ratio 7:1 for player A.

The following solution is based on a method described in the correspondence between Pascal and Fermat.[2] In the given situation at most three games are played. Assume that the players will play exactly three games, although in most cases the winner will be known sooner. Denote winning games a and b by players A and B, respectively. Then each of the following eight sequences has the same probability:

$$\begin{array}{cccc} aaa & aba & abb & bba \\ aab & baa & bab & bbb \end{array}$$

Only the last of them gives the prize to B: in all remaining cases A is the winner. This means that the fair division of the prize is really 7:1.

The following example suggests that the essence of probabilistic problems was also difficult for prominent mathematicians. Jean D'Alembert (1717–1783) calculated the probability that a head appears at least once in two tossings of a fair coin. He concluded that this probability is $\frac{2}{3}$ because there are three outcomes ($2\times$ head, $1\times$ head, $0\times$ head) and two of them are favorable. He published this solution in his *Encyclopedia* in 1754. Even Gottfried Leibniz (1646–1716), one of the founders of calculus, presented a wrong solution. The correct result is $\frac{3}{4}$, because there are four equally probable outcomes (head — head, head — tail, tail — head, tail — tail), three of which are favorable.

Much time elapsed until an appropriate mathematical model of a random experiment was constructed. The model developed by Andrey Kolmogorov (1903–1987) is now used in the most cases.

Assume that we have a set Ω, the elements of which are denoted by ω. Different elements are distinguished by specific indices. Elements ω are called *elementary events*, and Ω is the *space of elementary events*. If we want to describe rolling dice, the number of points that can appear on the upper face of the dice can be considered elementary events. Briefly we write $\omega_1 = 1, \dots, \omega_6 = 6$.

It is further assumed that a system $\mathcal{A}$ of subsets of Ω is given such that it is a σ-algebra. It means that

1. The system $\mathcal{A}$ is not empty.

[1] Probably S. del Ferro, round 1515, was the first to find the solution of the cubic equation in radicals but he did not publish it. Independently, N. Fontana (nicknamed Tartaglia) derived the same result in 1535. He kept his solution secret untill G. Cardano got it from him and published it in his work *Ars Magna*. A solution of the fourth-degree equation was found by L. Ferrary, who was an assistant of Cardano. See Rosen (1995).

[2] Székely (1986, p. 11), and Bühlman (1998, p. 163), write that it was Fermat who discovered the solution presented here. On the other hand, Mačák (1997, p. 14), claims that the solution was proposed in Pascal's letter to Fermat dated on August 24th, 1654. Except for Pascal's first letter (which disappeared), the correspondence between Pascal and Fermat was published in Smith (1929, pp. 546–565.)

2. If $A \in \mathcal{A}$, then the complement $A^c \in \mathcal{A}$.

3. If $A_1, A_2, \cdots \in \mathcal{A}$, then $\bigcup_{i=1}^{\infty} A_i \in \mathcal{A}$.

Random events (or briefly events) are sets belonging to $\mathcal{A}$. If A_i are events, then De Morgan's (1806–1871) formula $\cap_{i=1}^{\infty} A_i = (\cup_{i=1}^{\infty} A_i^c)^c$ implies that $\cap_{i=1}^{\infty} A_i$ is also an event. Since the system $\mathcal{A}$ is not empty, it contains an event A. Then it also contains A^c. So $\mathcal{A}$ contains the *impossible event* $\emptyset = A \cap A^c$ as well as the *sure event* $\Omega = \emptyset^c$.

If the space Ω contains only finite or countable infinite many elements, $\mathcal{A}$ is always the system of all subsets of Ω. If Ω is not countable, $\mathcal{A}$ usually contains only *some* of its subsets. Finally, it is assumed that a *probability measure* P is defined on the sets $A \in \mathcal{A}$. It fulfills following assumptions:

1. **Nonnegativity:** $\mathsf{P}(A) \geq 0$ for every set $A \in \mathcal{A}$.

2. **σ-additivity:** If $A_1, A_2, \ldots$ are disjoint sets belonging to $\mathcal{A}$, then we have $\mathsf{P}\left(\bigcup_{i=1}^{\infty} A_i\right) = \sum_{i=1}^{\infty} \mathsf{P}(A_i)$.

3. **Normalization:** $\mathsf{P}(\Omega) = 1$.

Here we can see why we carefully assumed that $\mathcal{A}$ need not contain all subsets of Ω. It is known that in the case $\Omega = [0, 1]$ it is not possible to define a measure P on all its subsets in such a way that $\mathsf{P}([a, b]) = b - a$ for all $0 \leq a < b \leq 1$ (e.g., see Natanson 1957). As a rule, we consider in this case only the Borel σ-algebra — the smallest σ-algebra containing all the intervals $[a, b]$ mentioned. The *probability* of an event $A \in \mathcal{A}$ is the value $\mathsf{P}(A)$.

Assumptions 1. – 3. imply some elementary consequences. Let A be an event and A^c be the *complement* to A. Since $A \cup A^c = \Omega$ and the events A, A^c are *disjoint* (*mutually exclusive*), assumptions 2 and 3 yield $\mathsf{P}(A^c) = 1 - \mathsf{P}(A)$. Since $\mathsf{P}(\Omega) = 1$ and $\emptyset$, Ω are disjoint, we have $\mathsf{P}(\emptyset) = 0$ from assumption 2. On the other hand, the relation $\mathsf{P}(A) = 0$ does not imply $A = \emptyset$. The triple $(\Omega, \mathcal{A}, \mathsf{P})$ is called *probability space*. This general model does not solve the problem, of how to choose Ω, $\mathcal{A}$, and P for a given random experiment. We describe this later.

Originally, probability was introduced in a different way. Property 2 formulated for only a finite number of disjoint events $A_1, \ldots, A_n$ was an assertion known as the *theorem of addition of probabilities*.

1.2 CLASSICAL PROBABILITY

If Ω has only a finite number of elements, and preliminary considerations ensure that all elementary events are equally probable, then the calculation of probability of an event might be easy. Assume that $\Omega = \{\omega_1, \ldots, \omega_n\}$ and $\mathsf{P}(\{\omega_1\}) = \cdots = \mathsf{P}(\{\omega_n\})$. Then properties 2 and 3 yield $\mathsf{P}(\{\omega_i\}) = 1/n$ for $i = 1, \ldots, n$. For instance, if we buy a dice, the manufacturer assures that it is homogeneous. Then we deduce that rolling it normally, each of numbers 1,. . . ,6 has the same probability, namely, $\frac{1}{6}$, to appear. If we want to calculate the probability of a general event, we must first calculate the number of elementary events of which it consists. These are *favorable cases*. According to assumption 2, probability of the event considered is the ratio of the number of favorable cases to the number of all elements of the space Ω. This method is demonstrated in several examples.

One of the most popular lotteries in the Czech Republic is *Sportka*. The player guesses 6 of 49 numbers. If all six numbers are guessed correctly, the player wins the first prize. What is the probability of this event? It is guaranteed that all six numbers will appear with the same probability. The space Ω can be considered as space of all subsets with six elements within $\{1, \ldots, 49\}$. We have $\binom{49}{6}$ such subsets. There is only one favorable subset corresponding to the first prize, and so the probability that the player wins the first prize is

$$\frac{1}{\binom{49}{6}} = \frac{1}{13,983,816}.$$

In another example, a player gets five playing cards from a shuffled set of standard playing cards. A poker player, for instance, might be interested in knowing the probability that four of those cards are aces. Five cards from 32 can be chosen $\binom{32}{5}$ ways. If the cards are well shuffled, then any five cards have the same probability. The number of favorable cases is 28, because any of the remaining 28 cards can be chosen to complete the set of four aces. So the probability, which we are looking for, is

$$\frac{28}{\binom{32}{5}} = \frac{28}{201,376} = 0.000139.$$

If you play poker and in the first round you get four aces more frequently than once in $10,000$ plays, the cards may not be shuffled well. Or perhaps there is another reason for this coincidence.

Sometimes there are several possibilities for defining the space of elementary events Ω to a given random experiment. If this is done correctly, we get different procedures leading to the same true result. The following text shows that even professionals may not find the simplest method.

Ridenhour and Woodward (1984) write that the popular fast-food chain McDonald's sponsored a play called "Star Raiders." Every lot had 10 cells. In 2 cells there was a picture of the prize that the participant could win. The prize ranged from a hamburger to a personal computer. The other two cells contained the inscription ZAP. The remaining 6 cells were empty. The owner of the lot erased individual cells successively. If she discovered both symbols of the prize before any ZAP, she got the prize. In the reverse case she got nothing.

Ridenhour and Woodward calculated the probability of winning as follows. Let M be a cell containing a picture of the prize from McDonald's, Z a cell with ZAP, and N an empty cell. Two cells for M can be chosen in $\binom{10}{2}$ ways. If we fix them, then we can choose cells for Z in $\binom{8}{2}$ ways. These possibilities can be combined. So the number of ways to distribute cells for M, Z, and N is

$$n = \binom{10}{2}\binom{8}{2} = 1260.$$

Lots were printed and sold so that all orderings of symbols were equally probable. To calculate the number of orderings where both M precede both Z, we consider the following cases.

1. The first Z is in the third cell. Then M must be in the first and second cell, the second Z can be in any of remaining 7 cells. We have $\binom{2}{2}\binom{7}{1}$ orderings.

2. The first Z is in the fourth cell. Both M precede, the other Z is in one of last 6 cells. Number of such orderings is $\binom{3}{2}\binom{6}{1}$.

We continue similarly. Finally we come to the situation where the first Z is in the penultimate cell. In such a case we have $\binom{8}{2}\binom{1}{1}$ orderings.

Total number of all favorable orderings is

$$\binom{2}{2}\binom{7}{1} + \binom{3}{2}\binom{6}{1} + \binom{4}{2}\binom{5}{1} + \binom{5}{2}\binom{4}{1} + \binom{6}{2}\binom{3}{1} + \binom{7}{2}\binom{2}{1} + \binom{8}{2}\binom{1}{1} = 210.$$

The probability of winning the prize is $P = \frac{210}{1260} = \frac{1}{6}$.

In the general case we could consider a lot having m cells with a prize, z cells with ZAP, and k empty cells. To win, it is necessary to find all m cells with prize before any cell with ZAP. Here the number of all orderings is

$$n = \binom{m+z+k}{m}\binom{z+k}{z}$$

and the number of favorable orderings is

$$\binom{m}{m}\binom{k+z-1}{z-1} + \binom{m+1}{m}\binom{k+z-2}{z-1} + \cdots + \binom{m+k}{m}\binom{z-1}{z-1} = \binom{m+z+k}{k}$$

(see Kaucký 1975, p. 48). This result can also be derived by the following method. Empty cells can be chosen in $\binom{m+z+k}{k}$ ways. Only one ordering of the cells with M and Z leads to a win. So the number of winning orderings is $\binom{m+z+k}{k}$, and the probability of winning is

$$P = \frac{\binom{m+z+k}{k}}{\binom{m+z+k}{m}\binom{z+k}{z}}. \tag{1.1}$$

In the special case described above we had $m = z = 2$, $k = 6$.

This reasoning can be simplified (Randall 1984). It is not necessary to consider empty cells. A choice of such a cell represents only a short delay before selecting a cell containing M or Z. If $m = z = 2$, then four cells containing M or Z can be ordered in $\binom{m+z}{m} = \binom{4}{2} = 6$ ways. Only one of these orderings produces a win, namely, $MMZZ$. So we have $P = \frac{1}{6}$.

In the general case the number of orderings of symbols M and Z is $\binom{m+z}{m}$ and only one of them wins — such where all M are at beginning. Therefore

$$P = \frac{1}{\binom{m+z}{m}}.$$

Note that formula (1.1) after some simplification gives the same result.

Now, we consider the *problem about shoes thrown into a box* (see Gordon 1997, p. 70, Exercise 50). Six different pairs of shoes are thrown in a big box. In the dark, five shoes are randomly drawn from the box. Find the probability p that there is at least one complete pair among the shoes selected.

There are $N = \binom{12}{5}$ ways to choose 5 shoes from a total of 12. The number of cases where each of the 5 shoes belongs to a different pair is $k = (12 \times 10 \times 8 \times 6 \times 4)/5!$. So the probability p is

$$p = 1 - \frac{k}{N} = \frac{25}{33} = 0.75758.$$

We further introduce the *problem about defective cameras* (see Gordon 1997, p. 70, Exercise 52). In a camera factory an inspector carefully checked a series containing 20 pieces and found that 3 of them needed adjustment before processing. A worker inadvertently returned three defective cameras among the others, so the order was random. The inspector had to recheck the cameras to find the three defective pieces. Answer the following questions.

1. Find the probability r_{17} that the inspector must check 17 cameras maximum.
2. Find the probability p'_{17} that he must check exactly 17 cameras.
3. Calculate the most probable number of the cameras that must be checked.

Three defective cameras can be placed among 17 good ones in $N = \binom{20}{3}$ ways. Three defective pieces can be placed on the first 17 positions in $k = \binom{17}{3}$ ways. The probability that 3 defective cameras are among the first 17 checked is

$$r^*_{17} = \frac{k}{N} = \frac{680}{1140} = 0.596491.$$

If the inspector finds only good cameras among the first 17 places, it clearly is unnecesssary to check the remaining three pieces. The probability that three defective cameras are at the last three places is

$$r^{**}_{17} = \frac{1}{N} = \frac{1}{1140} = 0.000877.$$

Since both possibilities are disjoint, we get

$$r_{17} = r_{17}^* + r_{17}^{**} = 0.597368.$$

The number of orderings such that a defective camera is in the 17th place and the remaining two are somewhere in the preceding 16 places is $m_{17} = \binom{16}{2}$. The probability of this event is

$$p_{17} = \frac{m_{17}}{N} = \frac{720}{6840} = 0.105263.$$

The probability that the inspector must check exactly 17 cameras is

$$p'_{17} = p_{17} + r_{17}^{**} = 0.106140.$$

The number of orderings such that the last defective camera is in the sth place is $m_s = \binom{s-1}{2}$. The probability that the last defective camera is in the sth place is

$$p_s = \frac{m_s}{N} = \frac{\binom{s-1}{2}}{\binom{20}{3}} \qquad k = 3, 4, \ldots, 20.$$

Using the known formula

$$\sum_{i=t}^{n} \binom{i}{t} = \binom{n+1}{t+1}, \qquad n \geq t,$$

we obtain $p_3 + \cdots + p_{20} = 1$. Obviously, $p_3 < p_4 < \cdots < p_{20}$. With the highest probability, the last defective camera will be found in the last place. This result could be called *Murphy's law.* Now, we calculate in detail the probability that the inspector must check exactly k cameras.

- If $3 \leq k \leq 16$, the inspector checks exactly k cameras if and only if two defective pieces are among the first $k - 1$ cameras and the kth camera is also defective.
- The inspector checks exactly 17 cameras if and only if two defective pieces are among the first 16 cameras and the 17th is also defective, or if all the first 17 cameras are good.
- The inspector checks exactly 18 cameras if and only if two defective pieces are among the first 17 cameras and the 18th is also defective, or if one defective camera is among the first 17 pieces and 18th is good.
- The inspector checks exactly 19 cameras if and only if two defective cameras are among the first 18 pieces.
- The inspector need not check all 20 cameras.

Let p'_k be the probability that the inspector must check exactly k cameras. It is clear that $p'_k = p_k$ for $3 \le k \le 16$. The value p'_{17} has already been calculated. Further, we get

$$p'_{18} = p_{18} + \frac{17}{\binom{20}{3}} = 0.106140, \qquad p'_{19} = 2 \times p_{19} = 0.268420, \qquad p'_{20} = 0.$$

With the highest probability, it will be necessary to check 19 cameras.

1.3 GEOMETRIC PROBABILITY

Sometimes the space Ω is a Borel subset of the Euclidean space $\mathbb{R}_n$ with a positive and finite Lebesgue measure $\lambda(\Omega)$. If it is ensured that $\mathsf{P}(A) = \lambda(A)/\lambda(\Omega)$, for every Borel set $A \subset \Omega$, then no elementary event $\omega \in \Omega$ is preferred and the probability of any Borel set $A \subset \Omega$ is proportional only to its Lebesgue measure. The measure P, called *geometric probability*, was applied in calculations for different experiments performed as early as the eighteenth century. At that time geometric probability was the only way to extend the calculation of the classical probability to some noncountable spaces Ω.

Consider an idealized watch with a classical face such that the second hand moves continuously. If we want to look at it, we can ask about the probability that the second hand points to a place between 1 and 3. This ark A forms one-sixth of the whole circle Ω. So the abovementioned probability is $\frac{1}{6}$.

The *problem of meeting* is usually solved by means of geometric probability. Two friends, say, X and Y, agreed to meet at a given place between noon and 1 p.m. Say that X comes to the appointed meeting place randomly during this interval and independently of Y, who also comes randomly. Each of them will wait for 10 minutes, but not after 1 p.m. Calculate the probability that the friends meet.

This situation is described in Fig. 1.1. The moment of arrival of Mr. X is represented as a randomly chosen point in segment AB and similarly the moment of arrival of Mr. Y a randomly chosen point in segment AD. The scale on both axes is considered in minutes. The friends meet if and only if the point (x, y) falls into polygon $AEFCGH$. The area of square $ABCD$ is 60^2. The sum of the areas of triangles EBF and GDH is 50^2. Then the area of the polygon $AEFCGH$ makes $60^2 - 50^2 = 1100$. The probability that X and Y meet is

$$P = \frac{1\,100}{60^2} = 0.306.$$

One problem solved using geometric probabilities showed strange results. A circle K is given and a chord is randomly chosen in it. Calculate the probability that the chord is longer than a side of an equilateral triangle inscribed in the circle K.

The first solution. We randomly choose a point X inside the circle, and this point will be the midpoint of the chord (see Fig. 1.2). Let k be the circle inscribed in the equilateral triangle ABC inscribed in the circle K (see Fig. 1.3). If X falls into the

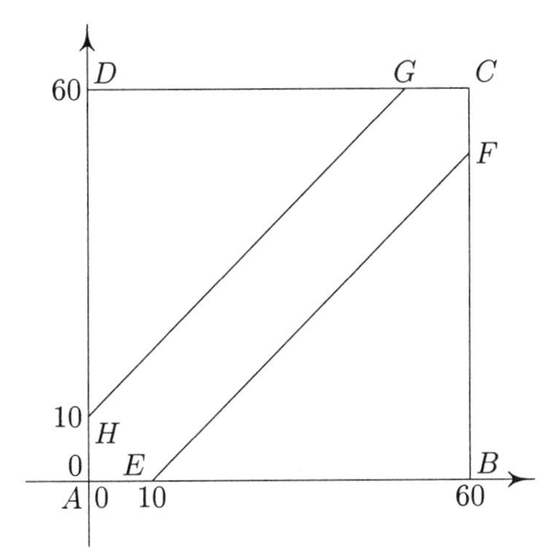

Fig. 1.1 Problem of meeting.

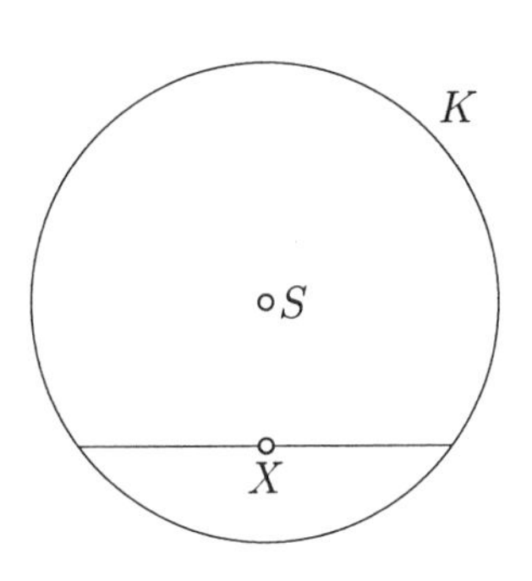

Fig. 1.2 Random chord.

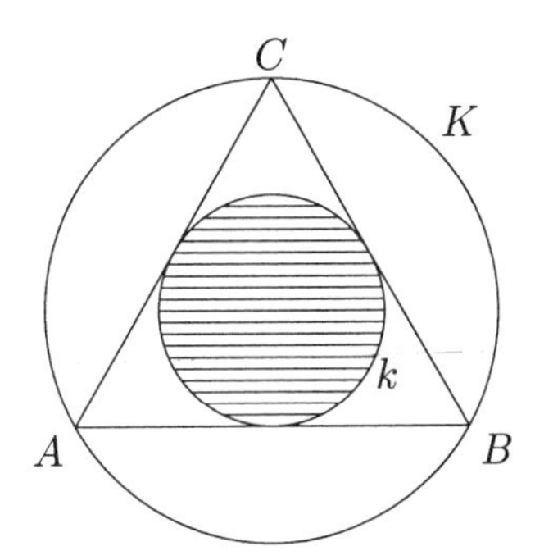

Fig. 1.3 First solution.

circle K, the chord is longer than a side of the triangle ABC. If X falls outside, the chord is shorter. If the circle K has radius R, then the radius r of the circle k inscribed in the triangle ABC is $r = R \sin 30° = R/2$ (see Fig. 1.4). The area of the circle k is $\pi r^2 = \pi R^2/4$, the area of the circle K is πR^2. Since the point X was chosen randomly in the circle K, the calculated probability is

$$P_1 = \frac{\pi R^2/4}{\pi R^2} = \frac{1}{4}.$$

The second solution. The length of the chord is determined by distance of its midpoint X from the center S of the circle K. Because of symmetry we can restrict our considerations to the case when the midpoint X is in the segment SY (see Fig. 1.4). Let U be the midpoint of the side AB. The chord is longer than the side AB if its midpoint X falls in the segment SU. We calculated above that the length of the segment SU is $r = R/2$. Since X is chosen randomly anywhere in the chord

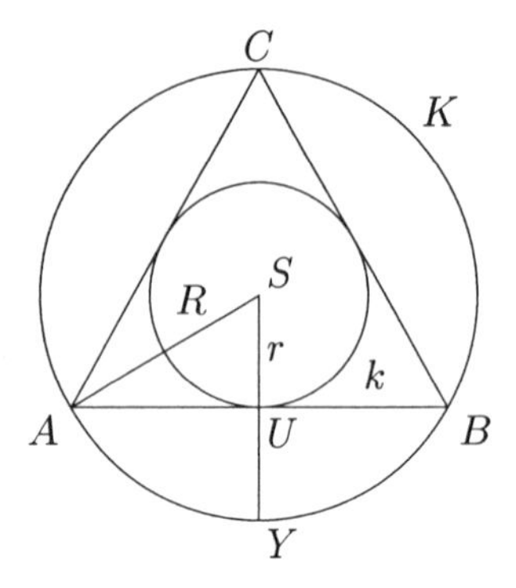

Fig. 1.4 Second solution.

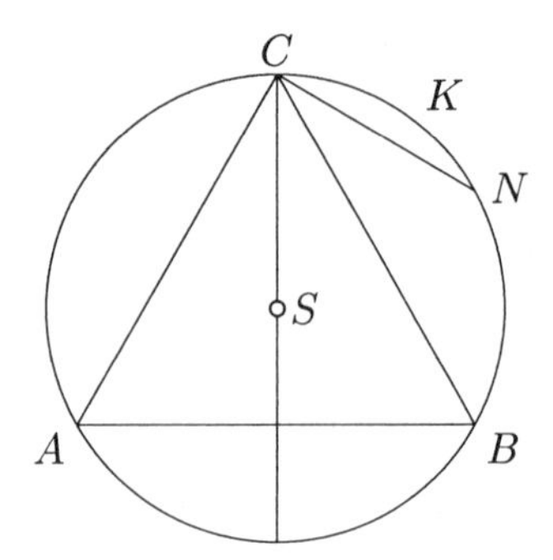

Fig. 1.5 Third solution.

SY, the probability is

$$P_2 = \frac{R/2}{R} = \frac{1}{2}.$$

The third solution. Assume that an endpoint of the chord is fixed, for example, that it is identical with the point C (see Fig. 1.5). The angle between the chord CN and the segment CS can reach arbitrary value between 0° and 90°. Because of symmetry we can restrict ourselves to the right half of the circle. The chord CN is longer than the side CB if the angle NCS is between 0° and 30°. Because this angle is chosen randomly from the interval 0° and 90°, the probability is

$$P_3 = \frac{30}{90} = \frac{1}{3}.$$

This example, known as *Bertrand's paradox*, was published in 1889.

The variation in different results seemed to be due to the fact that the set of points inside the circle K is not countable. But the explanation is much simpler. Each solution describes a different experiment with a different definition of a "randomly chosen" chord in the circle. It is necessary to know in advance which mechanism is involved in placement of the chord. This mechanism might be different from the three methods mentioned above, and the correct solution might be neither of the three solutions that we derived.

1.4 DEPENDENCE AND INDEPENDENCE

Let $A, B \in \mathcal{A}$ be two random events such that $\mathsf{P}(B) > 0$. The conditional probability of event A given that event B has occurred is defined by the formula

$$\mathsf{P}(A|B) = \frac{\mathsf{P}(A \cap B)}{\mathsf{P}(B)}. \tag{1.2}$$

The motivation leading to this definition can be found in most textbooks on probability theory.

If knowledge of the fact that event B has occurred has no influence on the probability of event A, we can say that event A does not depend on event B. It can be described by formula

$$\mathsf{P}(A|B) = \mathsf{P}(A). \tag{1.3}$$

If we insert terms from (1.2), then we can see that independence A on B yields

$$\mathsf{P}(A \cap B) = \mathsf{P}(A)\mathsf{P}(B). \tag{1.4}$$

Assume for a while that $\mathsf{P}(A) > 0$. Dividing both sides of the formula (1.4) by $\mathsf{P}(A)$, we get $\mathsf{P}(B|A) = \mathsf{P}(B)$. This means that if event A does not depend on event B, then B does not depend on A. Some time ago, independence of event A on event B has been defined by formula (1.3). Assertion (1.4) was called the *theorem of multiplication of probabilities*: If events A and B are independent, then the probability that they occur simultaneously is product $\mathsf{P}(A)\mathsf{P}(B)$ of probabilities of individual events. Today we take (1.4) as the definition of independence. An advantage of this definition is that we need not assume that any of probabilities $\mathsf{P}(A)$ or $\mathsf{P}(B)$ is positive and that the definition is symmetric with respect to A and B.

Theorem 1.1 *If A and B are independent events, then A and B^c, A^c and B, A^c and B^c are also pairs of independent events.*

Proof. Let A and B be independent events. Then (1.4) yields

$$\mathsf{P}(A \cap B^c) = \mathsf{P}(A) - \mathsf{P}(A \cap B) = \mathsf{P}(A) - \mathsf{P}(A)\mathsf{P}(B) = \mathsf{P}(A)\mathsf{P}(B^c).$$

The remaining two assertions can be proved analogously. □

Consider events $A_1, \dots, A_n$. We say that these events are *independent* if

$$\mathsf{P}(A_{i_1} \cap \dots \cap A_{i_k}) = \mathsf{P}(A_{i_1}) \times \dots \times \mathsf{P}(A_{i_k})$$

holds for each subset $\{i_1, \dots, i_k\}$ of the set $\{1, \dots, n\}$. Note that independent events satisfy $\mathsf{P}(A_1 \cap \dots \cap A_n) = \mathsf{P}(A_1) \times \dots \times \mathsf{P}(A_n)$. However, this relation itself does not imply that the events are independent, if $n > 2$.

To illustrate this point, we introduce the *problem concerning two inaccurate mathematicians*. Rabinowitz (1992, p. 288), writes that the problem was published in *Ontario Secondary School Mathematics Bulletin* **16** (1980/81, p. 18), and its solution, in the same journal (**16**, 1980/81, p. 19).

Aloysius and Bartholomew are two inaccurate mathematicians whose chances of solving a given problem correctly are 1 in 8 and 1 in 12, respectively. Assuming that they obtain the same result and the probability of them making the same error is 1 to 1001, find the probability that the result is correct.

Let C be the event that both boys reach the correct result and S the event that that both boys reach the same result. If N is the event that both boys reach the same incorrect result, then

$$\mathsf{P}(C) = \frac{1}{8} \times \frac{1}{12}, \qquad \mathsf{P}(N) = \frac{7}{8} \times \frac{11}{12} \times \frac{1}{1001}, \qquad \mathsf{P}(S) = \mathsf{P}(C) + \mathsf{P}(N),$$

and so

$$\mathsf{P}(C|S) = \frac{\mathsf{P}(C \cap S)}{\mathsf{P}(S)} = \frac{\mathsf{P}(C)}{\mathsf{P}(S)} = \frac{13}{14} = 0.929.$$

If Aloysius and Bartholomew obtain the same result, then this result is correct with probability 0.929.

Further we solve the *problem about the forgotten umbrella* (see Gordon 1997, p. 73, Exercise 62). An absent-minded professor of mathematics forgets his umbrella in a shop with probability $\frac{1}{4}$, under the condition, of course, that he entered the shop with an umbrella. The professor left his home with an umbrella, visited three shops and going home observed that he had no umbrella. Calculate the probability that he forgot his umbrella in the ith shop, $i = 1, 2, 3$.

Let A_i be the event that the professor forgets his umbrella in the ith shop. Then $A = A_1 \cup A_2 \cup A_3$ is the event that the professor forgets his umbrella in one of the three shops. Consecutively, we get

$$\mathsf{P}(A_1) = \frac{1}{4}, \qquad \mathsf{P}(A_2) = \frac{3}{4} \times \frac{1}{4}, \qquad \mathsf{P}(A_3) = \frac{3}{4} \times \frac{3}{4} \times \frac{1}{4}.$$

Since the events A_1, A_2, A_3 are disjoint, we have

$$\mathsf{P}(A) = \mathsf{P}(A_1) + \mathsf{P}(A_2) + \mathsf{P}(A_3) = \frac{37}{64}.$$

Since

$$\mathsf{P}(A_i|A) = \frac{\mathsf{P}(A_i \cap A)}{\mathsf{P}(A)} = \frac{\mathsf{P}(A_i)}{\mathsf{P}(A)},$$

we obtain

$$\mathsf{P}(A_1|A) = \frac{16}{37} = 0.4324, \qquad \mathsf{P}(A_2|A) = \frac{12}{37} = 0.3243,$$
$$\mathsf{P}(A_3|A) = \frac{9}{37} = 0.2432.$$

Our calculations show that the professor most likely forgot the umbrella in the first shop.

Another variant of this story is the *problem about forgotten luggage* (see Gordon 1997, p. 73, Exercise 63). A woman travels by airplain and is transported by three air companies. The probabilities that the luggage will be lost by the first, second, and third companies are 1%, 3%, and 2%, respectively. When the woman reached her destination, she observed that her luggage had disappeared. Calculate the probabilities that the luggage was lost by the first, second, and third companies.

Let A_i be the event that the luggage was lost during transportation by the ith company. Define $A = A_1 \cup A_2 \cup A_3$. We have

$$\begin{aligned} \mathsf{P}(A_1) &= 0.01, \\ \mathsf{P}(A_2) &= 0.99 \times 0.03 = 0.0297, \\ \mathsf{P}(A_3) &= 0.99 \times 0.97 \times 0.02 = 0.058906. \end{aligned}$$

Since

$$\mathsf{P}(A) = \mathsf{P}(A_1) + \mathsf{P}(A_2) + \mathsf{P}(A_3) = 0.058906$$

and $\mathsf{P}(A_i|A) = \mathsf{P}(A_i)/\mathsf{P}(A)$, we get

$$\mathsf{P}(A_1|A) = 0.170, \qquad \mathsf{P}(A_2|A) = 0.504, \qquad \mathsf{P}(A_3|A) = 0.326.$$

It is most likely that the luggage was lost by the second company.

1.5 BAYES' THEOREM

First, we formulate and prove the theorem of total probability. The theorem will be used not only in proof of Bayes' theorem but also in several applications (see Sections 2.8, 3.3, and 3.4).

Consider a finite or a countable infinite number of disjoint events $B_1, B_2, \dots$ such that one of them occurs (i.e., $B_i \cap B_j = \emptyset$ for each $i \neq j$ and $\cup_i B_i = \Omega$). Then $B_1, B_2, \dots$ is called the *complete system of events.* Sometimes it is claimed that these events form a *partition* of Ω.

Theorem 1.2 (of total probability) *Let $B_1, B_2, \dots$ be a complete system of events. If all probabilities $\mathsf{P}(B_m)$ are positive, the probability of an event A can be calculated from probabilities $\mathsf{P}(B_m)$ and from conditional probabilities $\mathsf{P}(A|B_m)$ using the formula*

$$\mathsf{P}(A) = \sum_m \mathsf{P}(A|B_m)\mathsf{P}(B_m). \tag{1.5}$$

Proof. Since

$$A = A \cap \Omega = A \cap (B_0 \cup B_1 \cup \dots) = (A \cap B_0) \cup (A \cap B_1) \cup \dots$$

and events $A \cap B_0, A \cap B_1, \dots$ are disjoint, we have

$$\mathsf{P}(A) = \sum_m \mathsf{P}(A \cap B_m).$$

We know that $\mathsf{P}(A \cap B_m) = \mathsf{P}(A|B_m)\mathsf{P}(B_m)$, and this concludes the proof. □

To illustrate the theorem of total probability, we present the *problem about an eccentric warden.* Rabinowitz (1992, p. 293), writes that the problem was published by K. M. Wilke in *The Pentagon* (**39**, 1980, p. 100), and a solution with some remarks was published in the same journal (**40**, 1981, pp. 111–112).

In a prison sits a prisoner who is sentenced to die. Fortunately, the warden, an eccentric person, offers the prisoner a chance to live. The warden gives the prisoner 12 black balls and 12 white balls. Next the warden gives the prisoner two boxes and tells him that tomorrow the executioner will randomly choose a box and will draw one ball at random from this box. If a white ball is drawn, the prisoner will be freed; if a black ball is drawn, the sentence will be carried out. How should the prisoner arrange the balls in the boxes so as to maximize his chance for freedom?

We denote the boxes by I and II. Assume that each of them must not be empty. The prisoner puts n balls into box I and i of them are white (say). Then box II contains $24 - n$ balls and $12 - i$ of them are white. The event A is that a white ball is drawn. Write $\mathsf{P}(A) = P(n, i)$ to emphasize that the probability $\mathsf{P}(A)$ depends on n and i. Let $\mathsf{P}(\text{I})$ and $\mathsf{P}(\text{II})$ be the probability that the executioner chooses boxes I and II, respectively. From the formulation of the problem we know that $\mathsf{P}(\text{I}) = \mathsf{P}(\text{II}) = \frac{1}{2}$. Applying theorem of total probability we obtain

$$\begin{aligned} P(n,i) &= \mathsf{P}(A|\text{I})\mathsf{P}(\text{I}) + \mathsf{P}(A|\text{II})\mathsf{P}(\text{II}) = \frac{i}{n}\mathsf{P}(\text{I}) + \frac{12-i}{24-n}\mathsf{P}(\text{II}) \\ &= \frac{1}{2}\left(\frac{i}{n} + \frac{12-i}{24-n}\right). \end{aligned}$$

Without loss of generality, assume that $n \le 12$ (otherwise we can change denotation of the boxes). It is easy to see that

$$P(12, i) = \frac{1}{2} \quad \text{for} \quad i = 0, \ldots, 12$$

and that

$$P(n, i) < P(n, n) \quad \text{for} \quad n = 1, \ldots, 11, \quad i = 0, \ldots, n-1.$$

Further, it can be verified that

$$P(n, n) < P(1, 1) \quad \text{for} \quad n = 2, \ldots, 12.$$

The inequalities imply that the prisoner maximizes his chance for freedom if he arranges balls in such a way that one box contains only one ball—namely white—and the other box contains all the remaining balls. This maximal probability is $P(1, 1) = 0.739$.

Another illustration is the *problem about weather* (see Stirzaker 1994, p. 51, Problem 43). Days are either sunny or cloudy. The weather tomorrow is the same as the weather today with probability p, or it is different with probability $q = 1 - p$. Let s_n be the probability that n days later the weather is the same as it is today. The probability s_n can be calculated from theorem of total probability if we take into account that the weather tomorrow can be either the same as or different from the weather today. This leads to the formula

$$s_n = p s_{n-1} + q(1 - s_{n-1}), \qquad n \ge 1,$$

where $s_0 = 1$. After some rearrangement we get

$$s_n = (p - q)s_{n-1} + q, \qquad n \ge 1,$$

and using complete induction, we obtain

$$s_n = \frac{1}{2}\left[1 + (p - q)^n\right].$$

Clearly, this model is oversimplified, and so it will be of little help for meteorologists.

Theorem 1.3 (Bayes' theorem) *Let A be a random event and $B_1, \dots, B_n$ a complete system of events. If $\mathsf{P}(A) > 0, \mathsf{P}(B_1) > 0, \dots, \mathsf{P}(B_n) > 0$, then*

$$\mathsf{P}(B_k|A) = \frac{\mathsf{P}(A|B_k)\mathsf{P}(B_k)}{\sum_{i=1}^{n} \mathsf{P}(A|B_i)\mathsf{P}(B_i)}, \qquad k = 1, \dots, n. \tag{1.6}$$

Proof. From the definition of conditional probability, we have

$$\mathsf{P}(B_k|A) = \frac{\mathsf{P}(A \cap B_k)}{\mathsf{P}(A)}.$$

We know that $\mathsf{P}(A \cap B_k) = \mathsf{P}(A|B_k)\mathsf{P}(B_k)$. The theorem of total probability gives

$$\mathsf{P}(A) = \mathsf{P}(A|B_1)\mathsf{P}(B_1) + \cdots + \mathsf{P}(A|B_n)\mathsf{P}(B_n).$$

Formula (1.6), which is called *Bayes' formula*, is proved. □

According to the inscription on his tomb in Bunhill Field cemetery in London, Thomas Bayes died on April 7, 1761, at age of 59. It seems likely that he was born in 1701. Bayes stated his result for a uniform prior, specifically for the case that all events B_k are equally probable. Stigler (1986) writes that it was Pierre - Simon Laplace (1749–1827) who apparently unaware of Bayes' work, stated the theorem in its general discrete form (1.6). Details can be found in Bernardo and Smith (1994, pp. 1–2).

1.6 MEDICAL DIAGNOSTICS

Formula (1.6) is very important. Quite frequently the so-called *prior probabilities* $\mathsf{P}(B_1), \dots, \mathsf{P}(B_n)$ are known. They can be probabilities of some causes or diseases. Conditional probabilities of an event A given that an event B_i has occurred may be known from experience, that is we may also know probabilities $\mathsf{P}(A|B_1), \dots, \mathsf{P}(A|B_n)$. The event A may be fever, cough, bleeding, or other symptoms. Formula (1.6) allows us to calculate *posterior probabilities* $\mathsf{P}(B_k|A)$ for each k, namely, the probabilities of individual diseases given the observed symptom.

We introduce an example, which can be found in many books (e.g., Gottinger 1980). Let C be the event that a randomly chosen person suffers from cancer. Then $Z = C^c$ is the event that the person does not suffer from cancer. Assume that a test for cancer has a reliability 0.95 in the following sense. If the person suffers from cancer then the test is positive with probability 0.95. This is called the *sensitivity* of the test. If the person does not suffer from cancer, the test will be negative also with probability 0.95 (this is called *specificity* of the test). We have chosen both probabilities as equal, although they are usually different. Define + and − positive and negative results of the test, respectively. Then

$$\mathsf{P}(+|C) = 0.95; \quad \mathsf{P}(-|C) = 0.05; \quad \mathsf{P}(+|Z) = 0.05; \quad \mathsf{P}(-|Z) = 0.95.$$

Table 1.1 Probability that a person suffers from cancer.

i	0	1	2	3	4	5
P_i	0.005	0.087	0.645	0.972	0.998	1.000

It is known that 0.5% of the given population suffers from cancer. This number represents the *prevalence* of the disease.[3] If we randomly choose a person as the subject and the test is positive, find the probability that the person really suffers from cancer.

Most people answer without hesitation that the person suffers from cancer with probability 0.95, because this is the reliability of the test. But this is not so. We know that $\mathsf{P}(C) = 0.005$. Bayes' theorem gives

$$\begin{aligned} P_1 = \mathsf{P}(C|+) &= \frac{\mathsf{P}(+|C)\mathsf{P}(C)}{\mathsf{P}(+|C)\mathsf{P}(C) + \mathsf{P}(+|Z)\mathsf{P}(Z)} \\ &= \frac{0.95 \times 0.005}{0.95 \times 0.005 + 0.05 \times 0.995} \\ &= 0.087. \end{aligned}$$

This surprising result can be explained to nonmathematicians (e.g., physicians) in the following way. If the test were applied to 100,000 persons then about 99,500 of them are healthy and about 500 suffer from cancer. The test gives a false-positive result in 5% healthy persons, so 4975 of them are declared to have cancer. From 500 ill persons, 95% will be detected to suffer from cancer, which equals 475 persons. Altogether we have $4975 + 475 = 5450$ positive tests, but only 475 of them belong to ill people. The ratio $475/5450$ is equal to 0.087.

Bayes' formula allows us to cumulate experience. If the test were positive for a given person, the person would suffer from cancer with probability 0.087. What happens if the test is applied again to this person? If the second test is independent of the first test, then the new probability P_2 can be calculated by the same formula as P_1, only instead of $\mathsf{P}(C)$, we have 0.087 and instead of $\mathsf{P}(Z)$, we insert 0.913. This gives $P_2 = 0.645$. Continuing in independent tests with positive results, we obtain probabilities given in Table 1.1.

If five tests give positive results, the probability is $P_5 = 0.999920$, which gives, after rounding off, the value 1.000 introduced in the table. However, it is necessary to note that the assumption of independence of tests may not be realistic.

To finish this solution on a more optimistic note, we calculate the probability Q_1 that a randomly chosen person does not suffer for cancer if the test has been negative.

[3]Sometimes the test is applied only to those persons who are not known to suffer from the disease. Then we consider the *incidence* of the disease, which is the number of new cases of the disease appearing in a given time interval (usually a year) among all inhabitants, or the number of new cases related to 1000 inhabitants.

Table 1.2 Percentage of positive responses.

	Test							
Disease	1	2	3	4	5	6	7	8
A	30	20	50	70	40	80	10	60
B	70	30	90	80	20	40	90	50
C	60	80	40	20	60	70	50	10

This time Bayes' theorem gives

$$\begin{aligned} Q_1 &= \mathsf{P}(Z|-) = \frac{\mathsf{P}(-|Z)\mathsf{P}(Z)}{\mathsf{P}(-|Z)\mathsf{P}(Z) + \mathsf{P}(-|C)\mathsf{P}(C)} \\ &= \frac{0.95 \times 0.995}{0.95 \times 0.995 + 0.05 \times 0.005} = 0.999\,736. \end{aligned}$$

This same problem is described in some books published in the 1990s (e.g., Wang 1993), only AIDS (autoimmunodeficiency syndrome) is considered instead of cancer. To get a clearer idea about real numbers, we present some data from Germany taken from Boer (1993). The incidence of AIDS was 0.1%, sensitivity of AIDS tests was 99.8%, and specificity of the tests was 99%.

Certain medical tests when administered alone do not lead to a correct diagnosis with a high probability, but when two or more medical tests are combined, it may be possible to make the correct inference.

Mead (1992) introduces the following example, which was developed from an idea presented by Aitchison during a seminar at Reading (California) around 1975. A preliminary diagnosis suggests that we have a patient with nontoxic goiter. There are three particular forms of this disease:

A: Simple goiter

B: Hashimoto's disease

C: Thyroid cancer

We denote these forms by A, B, and C, respectively. Eight tests are administered to enhance determination of diagnosis. None of them is perfect. The proportions of positive responses, determined from the results of the tests for patients observed previously, are given in Table 1.2.

If a patient suffers from disease A, then the first test will be positive with probability 0.3. If she suffers from B, the first test gives positive result with probability 0.7. Similarly we can read the remaining data in Table 1.2.

Assume for simplicity that all three forms of nontoxic goiter occur with the same frequency. Then their prior probabilities $\mathsf{P}(A)$, $\mathsf{P}(B)$, and $\mathsf{P}(C)$ will be equal:

$$\mathsf{P}(A) = \mathsf{P}(B) = \mathsf{P}(C) = \frac{1}{3}.$$

Table 1.3 Results of tests.

Number of test	1	2	3	4	5	6	7	8
Result of test	−	+	+	+	−	+	−	+

Table 1.4 Conditional probabilities of diseases A, B, and C after tests 1 to j.

j	1	2	3	4	5	6	7	8
Conditional probability of A	0.50	0.26	0.25	0.35	0.32	0.46	0.79	0.90
Conditional probability of B	0.21	0.16	0.29	0.46	0.57	0.40	0.08	0.07
Conditional probability of C	0.29	0.58	0.46	0.19	0.11	0.14	0.13	0.03

Suppose that the results of tests 1 to 8 (in this order) are those introduced in Table 1.3.

Let 1^- denote the event that the first test was negative. A similar denotation is used for the results of the remaining tests. Bayes' theorem gives

$$\begin{aligned}
\mathsf{P}(A|1^-) &= \frac{\mathsf{P}(1^-|A)\mathsf{P}(A)}{\mathsf{P}(1^-|A)\mathsf{P}(A)+\mathsf{P}(1^-|B)\mathsf{P}(B)+\mathsf{P}(1^-|C)\mathsf{P}(C)},\\
\mathsf{P}(B|1^-) &= \frac{\mathsf{P}(1^-|B)\mathsf{P}(B)}{\mathsf{P}(1^-|A)\mathsf{P}(A)+\mathsf{P}(1^-|B)\mathsf{P}(B)+\mathsf{P}(1^-|C)\mathsf{P}(C)},\\
\mathsf{P}(C|1^-) &= \frac{\mathsf{P}(1^-|C)\mathsf{P}(C)}{\mathsf{P}(1^-|A)\mathsf{P}(A)+\mathsf{P}(1^-|B)\mathsf{P}(B)+\mathsf{P}(1^-|C)\mathsf{P}(C)}.
\end{aligned}$$

This implies that

$$\mathsf{P}(A|1^-) = 0.50, \qquad \mathsf{P}(B|1^-) = 0.21, \qquad \mathsf{P}(C|1^-) = 0.29.$$

If the second test is independent of the first test, we obtain probabilities $\mathsf{P}(A|1^-,2^+)$ from analogous formulas, only $\mathsf{P}(A)$ must be substituted by $\mathsf{P}(A|1^-)$ and so on. The results are summarized in Table 1.4.

The diagnostic path for these results is shown in Fig. 1.6. Each triple of conditional probabilities is a point in three-dimensional space, but their sum equals to 1, and so a planar picture also suffices. The conditional probability of disease A is represented by distance from the segment BC toward point A. Other coordinates are used similarly.

When all tests are carried out, we should deduce that the patient suffers from disease A. Not all fairytales have a happy ending, nor does the hypothetical (based on a simulation experiment) case described here. It was known from the beginning that the patient suffers from disease B. We can see in Fig. 1.6 that the wrong diagnosis was caused mainly by the result of test 7. In the case of disease B, this test is positive in 90% of patients, but our patient belonged to the remaining 10%. To please you a little we should inform you that it did not concern a real patient but the result of a simulation experiment in a lesson at school.

Mead asks students to try to answer following questions, which we could also expect to be asked by a physician:

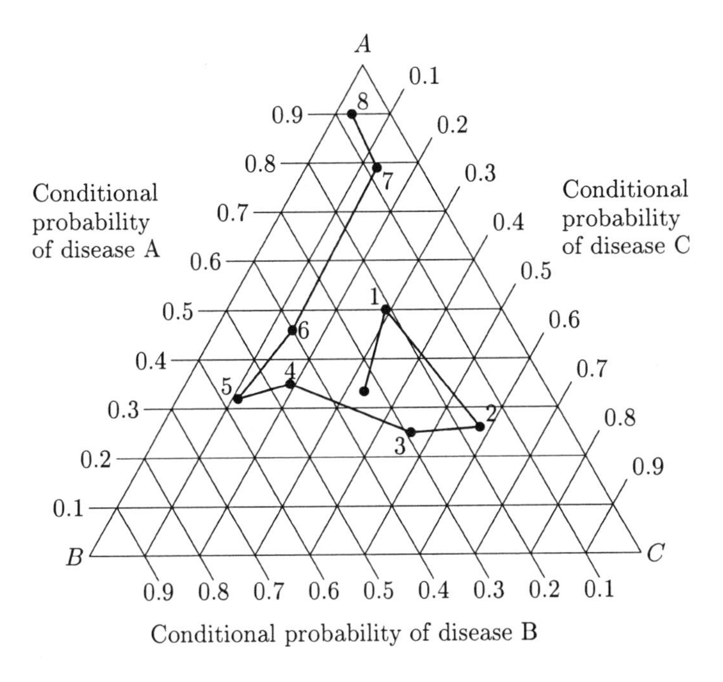

Fig. 1.6 Conditional probabilities of diseases A, B, and C.

1. Which tests should be used first?

2. Which tests should be administered if their costs are different and money is limited by a given amount?

3. Should the order in which tests are used depend on the previous results?

1.7 RANDOM VARIABLES

Assume that a real number corresponds to every element $\omega \in \Omega$. This means that we have a function defined on Ω with values in $\mathbb{R}$. In calculus functions are usually denoted f, g, and so on, but probability theory prefers capital letters from the end of the alphabet, eventually with some greek letters, for example ξ, η, and ζ. Let $X(\omega)$ be a number corresponding to the elementary event ω. If we want to have a comfortable work with the function $X : \Omega \to \mathbb{R}$, we must make some assumptions about it. Let $\mathcal{B}$ be the Borel σ-algebra on the real line $\mathbb{R}$. We say that the function X is measurable if $X^{-1}(B) \in \mathcal{A}$ holds for every set $B \in \mathcal{B}$. A measurable function is called a *random variable*.

For every real x, we define a function $F(x) = \mathsf{P}(\{\omega : X(\omega) < x\})$. The function F is called the *distribution function* of the random variable X. The following two cases occur almost exclusively.

1. Discrete case. Random variable X reaches only a finite or countably infinite number of values. Assume that X takes only values $x_1, x_2, \ldots$ and that $\mathsf{P}(\{\omega : X(\omega) = x_i\}) = p_i$, $i = 1, 2, \ldots$. We say that X has a *discrete distribution*. Its distribution function F is a step function: steps are at the points $x_1, x_2, \ldots$, and the step at x_i has value p_i, $i = 1, 2, \ldots$.

2. Continuous case. If there exists a function f such that $F(x) = \int_{-\infty}^{x} f(t)\,\mathrm{d}t$, then we say that X has a *continuous distribution* and that f is its *density*. If the distribution function F is known, we can calculate the density f by the formula $f(x) = F'(x)$.

A random variable X is usually described by some basic characteristics. One of these is the *expectation* $\mathsf{E}X$, which is defined by

$$\mathsf{E}X = \int_{\Omega} X(\omega)\,\mathrm{d}\mathsf{P}(\omega). \tag{1.7}$$

We say that $\mathsf{E}X$ exists if there exists the integral on the right-hand side of (1.7). In some practical situations, we know the distribution function. It can be proved that an equivalent expression for $\mathsf{E}X$ is

$$\mathsf{E}X = \int_{-\infty}^{\infty} x\,\mathrm{d}F(x), \tag{1.8}$$

where the right-hand side is Lebesgue–Stieltjes integral with respect to the measure generated by the distribution function F. If X has a discrete or a continuous distribution, the expectation is given by

$$\mathsf{E}X = \sum_i x_i p_i, \qquad \text{or} \qquad \mathsf{E}X = \int_{-\infty}^{\infty} x f(x)\,\mathrm{d}x. \tag{1.9}$$

Instead of $\mathsf{E}X$, we often write simply μ.

Sometimes it is necessary to calculate the expectation of a random variable Y, which is a function of the variable X, say, $Y(\omega) = g[X(\omega)]$. Using definition (1.7), we have

$$\mathsf{E}g(X) = \int_{\Omega} g[X(\omega)]\,\mathrm{d}\mathsf{P}(\omega).$$

It can be proved that for discrete and continuous distribution of X the formulas

$$\mathsf{E}g(X) = \sum_i g(x_i)p_i \qquad \text{and} \qquad \mathsf{E}g(X) = \int_{-\infty}^{\infty} g(x)f(x)\,\mathrm{d}x \tag{1.10}$$

hold, respectively.

Theorem 1.4 (of total expectation) *Let X be a discrete random variable taking values $x_1, \ldots, x_n$. Let $B_1, \ldots, B_m$ be a complete system of events such that $\mathsf{P}(B_j) > 0$ for $j = 1, \ldots, m$. Define conditional expectation of the random variable X, given B_j, by the formula*

$$\mathsf{E}(X|B_j) = \sum_{i=1}^{n} x_i \mathsf{P}(X = x_i|B_j).$$

Then

$$\mathsf{E}X = \sum_{j=1}^{m} \mathsf{E}(X|B_j)\mathsf{P}(B_j).$$

Proof. From theorem of total probability we get

$$\begin{aligned} \mathsf{E}X &= \sum_{i=1}^{n} x_i \mathsf{P}(X = x_i) = \sum_{i=1}^{n} x_i \sum_{j=1}^{m} \mathsf{P}(X = x_i|B_j)\mathsf{P}(B_j) \\ &= \sum_{j=1}^{m} \left[\sum_{i=1}^{n} x_i \mathsf{P}(X = x_i|B_j) \right] \mathsf{P}(B_j) \\ &= \sum_{j=1}^{m} \mathsf{E}(X|B_j)\mathsf{P}(B_j). \quad \square \end{aligned}$$

This theorem is frequently used for calculation of expectation from conditional expectations. We can say that total expectation is expectation of conditional expectations. It is clear that the theorem can be easily generalized to cases in which $n = \infty$ or $m = \infty$.

We frequently use positive random variables. Then we can ask the following question: Does there exist a positive random variable X such that

$$\mathsf{E}X > 1, \qquad \mathsf{E}\frac{1}{X} > 1?$$

Of course, yes. It suffices to choose

$$X = \begin{cases} \frac{1}{2} & \text{with probability } \frac{1}{2}, \\ 2 & \text{with probability } \frac{1}{2}. \end{cases}$$

It can be easily confirmed that $\mathsf{E}X = \mathsf{E}(1/X) = \frac{5}{4}$. Other elementary examples of this kind can be found in Deshpande (1992).

We often calculate the raw moment of the second order $\mu_2' = \mathsf{E}X^2$. It is used for definition of *variance* σ^2 of the random variable X:

$$\sigma^2 = \mu_2' - \mu^2. \tag{1.11}$$

Instead of σ^2, we may also write $\operatorname{var} X$. Equivalently, instead of (1.11), we have $\sigma^2 = \mathsf{E}(X - \mu)^2$.

If two random variables X and Y are given in the probability space $(\Omega, \mathcal{A}, \mathsf{P})$, we may define their *simultaneous distribution function* F by

$$F(x, y) = \mathsf{P}(\{\omega : X(\omega) < x, Y(\omega) < y\}).$$

Symbol ω is usually dropped, and we simply write $F(x, y) = \mathsf{P}(X < x, Y < y)$. The comma between symbols on the right-hand side of the last two formulas is a shorthand notation for intersection.

Let F_1 and F_2 be the distribution functions of random variables X and Y, respectively. Let F be their simultaneous distribution function. If $F(x, y) = F_1(x)F_2(y)$ holds for all reals x, y, then we say that the variables X and Y are *mutually independent*. If X and Y are mutually independent and their expectations are finite, it can be proved that

$$\mathsf{E}(XY) = (\mathsf{E}X)(\mathsf{E}Y).$$

If the simultaneous distribution function F can be written in the form

$$F(x, y) = \int_{-\infty}^{x} \int_{-\infty}^{y} f(u, v)\, du\, dv,$$

then we say that the random vector $(X, Y)'$ has a *continuous distribution* and f is its simultaneous density. It is clear that in this case

$$f(x, y) = \frac{\partial^2 F(x, y)}{\partial x \partial y}.$$

holds almost everywhere. The case of more random variables is analogous.

1.8 MANG KUNG DICE GAME

The term mang kung means blind men in Chinese. The Mang Kung dice game is an old Chinese game and it is still fairly popular in southern China and Hong Kong. The game is played with six dice such that the ith dice has five blank faces and the number i on the sixth face ($i = 1, \ldots, 6$). The blank face is counted as zero.

Typically, three players participate in this game. Each of them contributes 7 coins into a pool, so the pool contains $\xi = 21$ coins at the beginning of the game. The content ξ of the pool changes during the game according to the following rules.

1. The player starting the first game is chosen at random. We denote this player as A.

2. The player whose turn is up throws all six dice and counts the total face value X.

3. If $X < \xi$, the player takes X coins from the pool and passes the dice to the player sitting on his left.

4. If $X > \xi$, the player must add $X - \xi$ coins to the pool and then passes the dice to the player sitting on his left.

5. If $X = \xi$, the player wins the game. He receives all ξ coins from the pool, and each player has to pay him another ξ coins. The winner starts the next new game.

We can ask a lot of questions, for example

1. What is the probability $p_k = \mathsf{P}(X = k)$ for $k = 0, 1, \ldots, 21$?
2. Is it equally probable that each player will win the game? Or does the starter have an advantage?
3. Is the expected payoff the same for each player?
4. What is the expected duration in number of throws per game?

Direct calculation of p_k is cumbersome and difficult for most values of k. We present a method that allows an elegant derivation of these probabilities. Let D_i $(i = 1, \ldots, 6)$ be a random variable representing the face value that appeared on the ith dice. We have

$$D_i = \begin{cases} i & \text{with probability } \frac{1}{6}, \\ 0 & \text{with probability } \frac{5}{6}. \end{cases}$$

We can write $X = D_1 + \cdots + D_6$. We introduce the function

$$g(t) = \mathsf{E}t^X = \sum_{k=0}^{21} t^k p_k. \tag{1.12}$$

Since random variables $D_1, \ldots, D_6$ are independent, $g(t)$ can be also expressed in the form

$$g(t) = \mathsf{E}t^{D_1 + \cdots + D_6} = (\mathsf{E}t^{D_1}) \times \cdots \times (\mathsf{E}t^{D_6}).$$

From

$$\mathsf{E}t^{D_i} = t^i \times \frac{1}{6} + t^0 \times \frac{5}{6} = \frac{1}{6}(5 + t^i),$$

we get

$$g(t) = \frac{1}{6^6}(5 + t)(5 + t^2) \times \cdots \times (5 + t^6). \tag{1.13}$$

Comparing this result with (1.12), we can see that p_k is the coefficient by t^k in (1.13). For this reason $g(t)$ is called the *probability generating function.* Formula (1.12) yields

$$p_k = \frac{g^{(k)}(0)}{k!}. \tag{1.14}$$

Of course, we insert terms for $g(t)$ from the formula (1.13). Some program packages calculate derivatives very quickly and reliably.

One can see from (1.13) that probabilities p_k typically have a common denominator value of $6^6 = 46,656$. For this reason Table 1.5 contains only corresponding numerators.

Answers to other questions are known only from simulations. It was observed that each player will win the game with practically the same probability. But if the game is played only once, then the expected payoff for player A is 1.17, the payoff for A's left-hand neighbor is 0.01, and the payoff for the third player is -1.18. From this point of view the play is not fair. Because the game is played many times

Table 1.5 Values $p_k \times 6^6$.

k	$p_k \times 6^6$	k	$p_k \times 6^6$	k	$p_k \times 6^6$	k	$p_k \times 6^6$
0	15,625	6	4,500	12	425	17	30
1	3,125	7	2,000	13	300	18	30
2	3,125	8	1,500	14	200	19	5
3	3,750	9	1,625	15	180	20	5
4	3,750	10	1,025	16	55	21	1
5	4,375	11	1,025				

and the starter in further games is chosen randomly, the differences quickly become negligible. The average length of the game is approximately 20 throws.

However, the game is not limited to three players. If there are n players, the initial contribution to the pool from each of them is $21/n$ coins. If $21/n$ is not an integer, an appropriate multiple of it is taken. For example, when $n = 4$, the stake can be 21 coins. When $n = 6$, the players can decide whether each of them will pay 7 or 14 coins.

If you have enough coins, you can start to play the game. Mang Kung dice are produced and sold in Hong Kong, for example. In 1996 a set of six dice cost about 75p, excluding postage. Chan (1996) is willing to help you, if you contact him. His e-mail address is `ecscws@nus.sg`.

1.9 SOME DISCRETE DISTRIBUTIONS

Consider an experiment, the result of which can be the occurrence of a random event A. If A occurs, we say that it is success. If A does not occur, we denote it as failure. Let $\mathsf{P}(A) = p$. The experiment is repeated independently n times. Let X be the number of successes in these n experiments. We look for the probability that X equals a given number $k \in \{0, 1, \ldots, n\}$.

There are different possibilities for the occurrence of k successes (and, consequently, $n - k$ failures). One of them is that the first k experiments give only successes and the remaining failures. Because of independence of the experiments, the theorem regarding the multiplication of probabilities states that this result has probability

$$\underbrace{p \times \cdots \times p}_{k} \times \underbrace{(1-p) \times \cdots \times (1-p)}_{n-k} = p^k (1-p)^{n-k}.$$

If k successes are on other places, we also get probability $p^k(1-p)^{n-k}$ since some factors change their positions only. The number of possibilities, of how to choose k position for successes when n places are available is $\binom{n}{k}$. All these cases are disjoint. So the resulting probability must be a sum of $\binom{n}{k}$ equal numbers $p^k(1-p)^{n-k}$. This gives

$$\mathsf{P}(X = k) = \binom{n}{k} p^k (1-p)^{n-k}, \qquad k = 0, 1, \ldots, n. \tag{1.15}$$

These probabilities belong to the *binomial distribution*, which we denote by $\mathsf{Bi}(n, p)$. Simple calculations give $\mu = np$, $\sigma^2 = np(1-p)$. Usually we write q instead of $1 - p$.

We introduce a problem, the solution of which is based on binomial distribution. It was Sir Isaac Newton (1643–1727) who had to deal with this problem, although not very willingly, it seems (see Glickman 1991). The famous diarist Samuel Pepys (1633–1703) submitted the following excerpt to Newton along with an intermediary, John Smith:

The Question
A— has 6 dyes in a box with which he is to fling a six
B— has in another box 12 dyes with which he is to fling 2 sixes
C— has in another box 18 dyes with which he is to fling 3 sixes
Q— whether B and C have not as easy a taske as A at even luck?

It seems that Newton solved the problem with some hesitation. In his first reply to Pepys on November 26, 1693, he indicated that he examined the problem for the simpler case when A throws one dice and B, two. It appeared that A had some advantage. Then Newton formulated the problem more precisely, and he assumed that:

1. A, B, and C throw simultaneously.
2. To throw a 6 means to throw at least one 6.
3. A, B, and C throw just once.

In his second letter to Pepys dated December 16, 1693, Newton described his solution in some detail.

If we have n independent experiments and in each of them an event A occurs with probability p, then the total number X of occurrences of the event A is a random variable with binomial distribution $\mathsf{Bi}(n, p)$, for which we have formula (1.15). In our case n is the number of dice, $p = \frac{1}{6}$ and $1 - p = q = \frac{5}{6}$. Newton obtained

$$\begin{aligned}
\mathsf{P}(A) &= \sum_{k=1}^{6} \binom{6}{k} p^k q^{6-k} = 1 - \binom{6}{0} p^0 q^6 = 0.6651, \\
\mathsf{P}(B) &= \sum_{k=2}^{12} \binom{12}{k} p^k q^{12-k} = 1 - \binom{12}{0} p^0 q^{12} - \binom{12}{1} p^1 q^{11} = 0.6187, \\
\mathsf{P}(C) &= \sum_{k=3}^{18} \binom{18}{k} p^k q^{18-k} = 1 - \sum_{k=0}^{2} \binom{18}{k} p^k q^{18-k} = 0.5973.
\end{aligned}$$

Since $\mathsf{P}(A) > \mathsf{P}(B) > \mathsf{P}(C)$, the tasks of the players are not equally difficult.

Sometimes it happens that probability p in binomial distribution is small but number n of experiments is large. Investigate situation when $n \to \infty$ and $np \to \lambda$. For fixed k, we get

$$\begin{aligned}
\lim_{n\to\infty} \binom{n}{k} p^k (1-p)^{n-k} \\
&= \lim_{n\to\infty} \frac{n(n-1)\dots(n-k+1)}{k!} \frac{(np)^k}{n^k} \left(1-\frac{np}{n}\right)^{n-k} \\
&= \lim_{n\to\infty} \frac{1}{k!} \left(1-\frac{1}{n}\right)\left(1-\frac{2}{n}\right)\dots\left(1-\frac{k-1}{n}\right) \\
&\qquad \times (np)^k \left(1-\frac{np}{n}\right)^n \left(1-\frac{np}{n}\right)^{-k} \\
&= \frac{\lambda^k}{k!} e^{-\lambda}.
\end{aligned}$$

We derived probabilities $\mathsf{P}(X = k) = \frac{\lambda^k}{k!} e^{-\lambda}$, $k = 0, 1, 2, \dots$ of *Poisson distribution* $\mathsf{Po}(\lambda)$. (Siméon-Denis Poisson lived during 1781–1840.) Calculations give $\mu = \lambda$, $\sigma^2 = \lambda$.

1.10 SOME CONTINUOUS DISTRIBUTIONS

If a random variable X has the distribution function

$$F(x) = \begin{cases} 0 & \text{for} \quad x < 0, \\ x & \text{for} \quad x \in [0,1], \\ 1 & \text{for} \quad x > 1, \end{cases}$$

then we say that X has the *rectangular distribution* on the interval $[0, 1]$. This distribution is denoted by $\mathsf{R}(0, 1)$. Its density is $f(x) = 1$ for $x \in [0, 1]$ and it vanishes outside this interval. Expectation and variance of this distribution are $\mu = \frac{1}{2}$ and $\sigma^2 = \frac{1}{12}$, respectively.

In queueing theory the *exponential distribution* $\mathsf{Ex}(\lambda)$ is used quite often. Its density is $f(x) = \lambda^{-1} e^{-x/\lambda}$ for $x > 0$ and zero for negative x. Here $\lambda > 0$, is a parameter. The distribution function of this distribution is $F(x) = 0$ for $x < 0$ and $F(x) = 1 - e^{-x/\lambda}$ for $x \geq 0$. Expectation and variance of the exponential distribution are $\mu = \lambda$ and $\sigma^2 = \lambda^2$, respectively.

Exponential distribution has an interesting characteristic property. Assume that the lifetime of a device is a random variable X with exponential distribution. Briefly, we write $X \sim \mathsf{Ex}(\lambda)$. After some time we check the device and find that it works correctly. So we know that the event $A = \{\omega : X(\omega) \geq a\}$ occurred. Find the probability that the device will work at least x additional time units.

Remember that $\mathsf{P}(X < b) = 1 - e^{-b/\lambda}$ holds for every $b > 0$. For the complement, we get $\mathsf{P}(X \geq b) = 1 - \mathsf{P}(X < b) = e^{-b/\lambda}$. Let $B = \{\omega : X(\omega) \geq a + x\}$ be the event that the device will work from the beginning at least $a + x$ time units.

We are interested in the conditional probability $\mathsf{P}(B|A)$. Calculations give

$$\begin{aligned} \mathsf{P}(B|A) &= \frac{\mathsf{P}(A \cap B)}{\mathsf{P}(A)} = \frac{\mathsf{P}(X \geq a, X \geq a + x)}{\mathsf{P}(X \geq a)} = \frac{\mathsf{P}(X \geq a + x)}{\mathsf{P}(X \geq a)} \\ &= \frac{e^{-(a+x)/\lambda}}{e^{-a/\lambda}} = e^{-x/\lambda}. \end{aligned}$$

If we know that after time a the device worked correctly, probabilistic parameters of its further lifetime would be the same as if we were to replace at time a the device by another one that is completely new.

We have derived the fact that the exponential distribution satisfies $\mathsf{P}(X \geq a + x | X \geq a) = \mathsf{P}(X \geq x)$. For this reason we say that exponential distribution is *memoryless*.

Normal distribution is one of the most important distributions. We denote it as $\mathsf{N}(\mu, \sigma^2)$, and its density is

$$f(x; \mu, \sigma^2) = \frac{1}{\sqrt{2\pi}\sigma} \exp\left\{-\frac{(x-\mu)^2}{2\sigma^2}\right\}, \qquad -\infty < x < \infty. \tag{1.16}$$

The parameters of this distribution are $\mu \in \mathbb{R}$ and $\sigma > 0$. Their denotation is chosen in such a way that the expectation of a random variable with the distribution $\mathsf{N}(\mu, \sigma^2)$ is just μ and its variance σ^2. This distribution is also called *Gaussian*. (Carl Gauss lived during 1777–1855.) The *standard normal distribution* is $\mathsf{N}(0, 1)$, and its density is

$$\varphi(x) = \frac{1}{\sqrt{2\pi}} \exp\left\{-\frac{x^2}{2}\right\}.$$

The distribution function of the standardized normal distribution

$$\Phi(x) = \int_{-\infty}^{x} \varphi(t)\,\mathrm{d}t$$

is one of the special functions. It cannot be expressed in terms of elementary functions.

Normal distribution is very important, since it represents a limit form of some other distributions. Some results have been derived by Abraham de Moivre (1667–1754) and Laplace. Proofs of the following two theorems can be found in Rényi (1970). They express the asymptotic behavior of binomial distribution $\mathsf{Bi}(n, p)$. [Incidentally, Alfréd Rényi (1921–1970) was a famous Hungarian mathematician and we introduce some of his results later on.]

Theorem 1.5 (local Moivre–Laplace theorem) *Let $q = 1 - p$. Assume that $n \to \infty$, $k \to \infty$ in such a way that the expression $|k - np|/\sqrt{npq}$ remains bounded.*

Define function f by formula (1.16). Then

$$\lim_{n \to \infty} \frac{\binom{n}{k} p^k q^{n-k}}{f(k; np, npq)} = 1.$$

Theorem 1.6 (integral Moivre–Laplace theorem) *Let y be an arbitrary real number. Then*

$$\lim_{n \to \infty} \sum_{\left\{k:\, \frac{k-np}{\sqrt{npq}} < y\right\}} \binom{n}{k} p^k q^{n-k} = \Phi(y).$$

2

Random walk

2.1 GAMBLER'S RUIN

Consider two gamblers, say, A and B. They play a series of games. Assume that there are no ties and that the result of any game is determined randomly. Let $p \in (0, 1)$ be the probability that A wins a game. Then $q = 1 - p$ is the probability that A loses and B wins the game. Let the results of games be independent of each other. At the beginning, A has z coins and B has $a - z$ coins. To avoid trivial situations, we assume that $0 < z < a$. If A wins a game, then B gives A a coin. If A loses a game, A gives a coin to B. The gamblers stop at the moment when one of them has no coins and is thus ruined. Typical questions asked here are the following:

1. Find the probability that A will be ruined.

2. Find the probability that B will be ruined.

3. Find the expected number of games. Briefly, we mention the expected length of the play and we measure it by the number of games.

4. Find the probability that the play stops at the nth game.

The problem can be formulated alternatively as a *random walk* of a particle. The particle moves on the line. At time $t = 0$ the particle is at the point $z \in (0, a)$. At time $t = 1, 2, \ldots$, the particle moves one unit right or one unit left. The right movement has probability p and left movement has probability q. There are absorbing barriers at points 0 and a. If the particle reaches one of these points, it is absorbed and cannot move furter. This model is important not only for probability theory but also for its practical applications.

Let q_z be the probability that A will be ruined. After the first game A has either $z+1$ coins (with probability p) or $z-1$ coins (with probability q). If $1 < z < a-1$, then the theorem of total probability gives

$$q_z = pq_{z+1} + qq_{z-1}. \tag{2.1}$$

If we define

$$q_0 = 1, \qquad q_a = 0, \tag{2.2}$$

then formula (2.1) also holds for $z = 1$ and for $z = a - 1$. Assume first that $p \neq q$. Then a general solution of the difference equation (2.1) is

$$q_z = A + B\left(\frac{q}{p}\right)^z. \tag{2.3}$$

We calculate the constants A and B so that formula (2.3) also holds for $z = 0$ and for $z = a$: in other words, boundary conditions (2.2) are fulfilled. This leads to equations

$$A + B = 1, \qquad A + B\left(\frac{q}{p}\right)^a = 0.$$

Thus the final solution is

$$q_z = \frac{\left(\frac{q}{p}\right)^a - \left(\frac{q}{p}\right)^z}{\left(\frac{q}{p}\right)^a - 1}. \tag{2.4}$$

This is a unique solution.

Let p_z be the probability that gambler B will be ruined. Probability p_z can be derived quite analogously, or we can see that p_z follows from (2.4) when we insert q, p, and $a - z$ instead of p, q, and z, respectively. This gives

$$p_z = \frac{\left(\frac{p}{q}\right)^a - \left(\frac{p}{q}\right)^{a-z}}{\left(\frac{p}{q}\right)^a - 1}.$$

It is easy to check that $q_z + p_z = 1$ so that the play finishes in a finite number of steps with probability 1.

If $p = q = \frac{1}{2}$, then a general solution of the difference equation (2.1) is $q_z = A + Bz$. If we want to fulfill boundary conditions (2.2), we come to a system of equations $A = 1$, $A + Ba = 0$. We get

$$q_z = 1 - \frac{z}{a}. \tag{2.5}$$

Analogously, we get the probability that B will be ruined:

$$p_z = 1 - \frac{a - z}{a} = \frac{z}{a}.$$

Here we also have $q_z + p_z = 1$. In both cases p_z is also the probability that A wins and q_z is the probability that B wins.

Let D_z be the expected length of the play. The theorem of total expectation implies that for $0 < z < a$, we have

$$D_z = (1 + D_{z+1})p + (1 + D_{z-1})q,$$

so that

$$D_z = pD_{z+1} + qD_{z-1} + 1. \tag{2.6}$$

Here $D_0 = 0$, $D_a = 0$. Assume first that $p \neq q$. Now, (2.6) is a nonhomogeneous difference equation. It can be shown that a particular solution of it is $D_z = z/(q-p)$. Then the general solution of the equation (2.6) is

$$D_z = A + B\left(\frac{q}{p}\right)^z + \frac{z}{q-p}.$$

To fulfill conditions $D_0 = 0$ and $D_a = 0$, we have

$$A + B = 0, \qquad A + B\left(\frac{q}{p}\right)^a = -\frac{a}{q-p}.$$

From here we have the result

$$D_z = \frac{z}{q-p} - \frac{a}{q-p}\frac{1-\left(\frac{q}{p}\right)^z}{1-\left(\frac{q}{p}\right)^a}. \tag{2.7}$$

If $p = q = \frac{1}{2}$, then a particular solution of (2.6) is $-z^2$. A general solution is $D_z = A + Bz - z^2$. From boundary conditions $D_0 = 0$, $D_a = 0$ we obtain

$$D_z = z(a - z). \tag{2.8}$$

Numerical results show that the expected length of the play is longer than people think.

Let $P(n)$ be probability that a gambler will be ruined in the nth game. The derivation of this probability is quite complicated and can be found in Feller (1968). The resulting formula is

$$P(n) = a^{-1}2^n p^{(n-z)/2} q^{(n+z)/2} \sum_{\nu=1}^{a-1} \left(\cos^{n-1}\frac{\pi\nu}{a} \cdot \sin\frac{\pi\nu}{a} \cdot \sin\frac{\pi z\nu}{a}\right). \tag{2.9}$$

This formula was derived by Joseph-Louis Lagrange (1736–1813) and is published in many textbooks. It is not commonly known but has been discovered repeatedly.

The expected length D_z satisfies

$$D_z = \sum_{n=1}^{\infty} nP(n).$$

However, the formula for $P(n)$ is complicated, and it seems that this procedure is seldom used. The method introduced above is preferred.

We introduce a generalization of the random walk mentioned above. Let the probability that A wins in the moment when he has k coins be $p(k)$, that is, it depends on k. Define $q(k) = 1 - p(k)$. If A starts with z coins, then the probability that B will be ruined is

$$\frac{1 + \sum_{j=2}^{z} \frac{q(1)...q(j-1)}{p(1)...p(j-1)}}{1 + \sum_{j=2}^{n} \frac{q(1)...q(j-1)}{p(1)...p(j-1)}}$$

for $z = 1, \ldots, n$ where the sum in numerator is zero if $z = 1$. This result is derived in Palacios (1999).

2.2 AMERICAN ROULETTE

European roulette has one cell denoted as zero, which is neither red nor black. *American roulette* has two such cells, namely, 0 and 00. If the ball falls in such a cell, no gambler wins and bets are forfeited to the bank. Coyle and Wang (1993) compare two plays.

Play 1. A coin is tossed. A gambler bets for a head. (Of course, she can also choose a tail. Few people can tell the difference between a head and a tail. Ask a numismatist, and you will be surprised.) If the player wins, she gets her bet back and a coin as a prize. If she loses, she has one coin less.

Play 2. A gambler bets a coin on red in American roulette. The probability of winning in this case is $18/38 = 0.473684$.

Consider the following two situations:

(a) The gambler has 900 coins and will bet one coin in play 1 until she has a million (or until she is ruined).

(b) The gambler has 900 coins and will bet one coin in play 2 until she has a thousand (or until she is ruined).

Question: Which is more probable, succeeding in situation (a) or in situation (b)?

Sound reasoning suggests that it is more probable to earn 100 coins in situation (b) and to have thousand. What does sound calculation suggest?

In situation (a) we have $z = 900$ and $a = 1,000,000$; here $p = q = \frac{1}{2}$. Using (2.5) we calculate the probability that the gambler's ruin is $1 - z/a$. The probability that the gambler will have a million is $P_1 = z/a = 900/1,000,000 = 0.0009$. Incidentally, the expected number of games is $899,190,000$, which we obtain from (2.8).

In situation (b) where $z = 900$, $a = 1\,000$, and $p = \frac{18}{38}$ the probability of having a thousand is $1 - q_{900}$. Formula (2.4) indicates that this probability is $1 - q_{900}$. The

Table 2.1 Probabilities of reaching 1000 coins.

Strategy	Probability of reaching 1 000 coins
To bet 1 coin	0.00003
To bet 5 coins	0.124
To bet 20 coins	0.592
DS	0.88

expected length of the play is 17,009, which we get from (2.7). This is incredible, but

$$\frac{P_1}{P_2} = 33.884.$$

Reaching a million is 33 times more probable than reaching a thousand. Sound calculation has won over sound reasoning.

Betting a coin is not a suitable strategy. Remember that the expected number of games is nearly 10^9 in situation (a). If a game lasts only a second, we need, on average 28.5 years for this play: moreover, with a high probability, we are ruined at the end instead of having a million.

Dubins and Savage (1965) proposed a different strategy that considerably raises the probability of having 1000 coins. Their strategy (which we denote the *DS strategy*) recommends betting 100 coins for the first time. If the gambler loses, she must bet 200 coins the second time and so forth. The gambler bets the money needed for reaching 1000 coins in one game. If the gambler does not have much money, she bets all.

Let $h(9)$ be the probability that the gambler will have 1000 coins if she has 900 coins. Define $h(8)$ and so on similarly. The DS strategy gives

$$\begin{aligned} h(9) &= p + qh(8) \\ &= p + q[p + qh(6)] = p + pq + q^2 h(6) \\ &= p + pq + q^2[p + qh(2)] = p + pq + pq^2 + q^3 h(2) \\ &= p + pq + pq^2 + q^3[ph(4) + q \times 0] = p + pq + pq^2 + pq^3 h(4) \\ &= p + pq + pq^2 + pq^3[ph(8) + q \times 0] = p + pq + pq^2 + p^2 q^3 h(8). \end{aligned}$$

This yields

$$h(8) = \frac{p(1+q)}{1 - p^2 q^2},$$

so that

$$h(9) = p + q\frac{p(1+q)}{1 - p^2 q^2}.$$

Since $p = 0.478$, we get $h(9) = 0.88$. [A different derivation is introduced in Billingsley (1986, p. 104 and formula 7.31).] Probabilities of reaching 1000 coins for different strategies are compared in Table 2.1.

The expected value of gambler's prize useing the DB strategy is

$$100 \times 0.88 - 900 \times 0.12 = -20.$$

The expected value of gambler's prize on betting one coin is

$$100 \times 0.00003 - 900 \times 0.99997 = -899.97.$$

We calculate the expected number of games needed for decision when the DS strategy is used. Let $m(9)$ be this expectation when the gambler has 900 coins. Define $m(9)$ and so on, similarly. We have

$$\begin{aligned} m(9) &= 1 + qm(8) \\ &= 1 + q[1 + qm(6)] = 1 + q + q^2 m(6) \\ &= 1 + q + q^2[1 + qm(2)] = 1 + q + q^2 + q^3 m(2) \\ &= 1 + q + q^2 + q^3[1 + pm(4)] = 1 + q + q^2 + q^3 + pq^3 m(4) \\ &= 1 + q + q^2 + q^3 + pq^3[1 + pm(8)] \\ &= 1 + q + q^2 + q^3 + pq^3 + p^2 q^3 m(8). \end{aligned}$$

This gives

$$m(8) = \frac{1 + q + q^2 + pq^2}{1 - p^2 q^2}, \qquad m(9) = 1 + q\frac{1 + q + q^2 + pq^2}{1 - p^2 q^2}.$$

Inserting this result, we obtain $m(9) = 2.09$. A decision is made on average in the second game when using the DS strategy.

A few numerical illustrations are introduced in Anděl (1993). Consider situation (b) with European roulette, where the probability of winning on black is $18/37 = 0.486486$. Probability that a gambler with 900 coins will have 1000 coins is now $P_3 = 0.0044863217$ and $P_1/P_3 = 0.20061$. This agrees with intuition: reaching 1000 is 5 times more probable than reaching 1,000,000. And if we consider tossing a fair coin instead of any kind of roulette, then 1000 is reached with probability $P_4 = 0.9$ and the ratio of probabilities is $P_1/P_4 = 0.001$. This is the number that people guess when solving the original problem, if the small difference between the probabilities $p = \frac{18}{38}$ and $\frac{1}{2}$ is neglected.

2.3 A RELUCTANT RANDOM WALK

Consider a game in which players A and B use n cards labeled $1, 2, \ldots, n$. At each move, one of the numbers $1, 2, \ldots, n$ is chosen at random and the player with this card must give it to the other player. The game continues until one player has all the cards.

The game can be represented by a particle that moves on the real line at points $0, 1, \ldots, n$. If it is at point j, it corresponds to the case when player A has j cards. The transition probability from point j to point $j - 1$ is j/n and transition probability from point j to point $j+1$ is $(n-j)/n$. Absorbing barriers are at points 0 and n. First we calculate the expected value of the number of steps of the particle. It is the expected value of the number of games until a gambler is ruined.

Let D_j be the expected value of number of games, given that A starts with j cards. After the first game, he has either $j-1$ cards with probability j/n or $j+1$ cards with probability $(n-j)/n$. Theorem of total expectation gives a system of equations for D_j

$$D_j = 1 + \frac{j}{n} D_{j-1} + \frac{n-j}{n} D_{j+1}, \qquad 1 \le j \le n-1, \tag{2.10}$$

with boundary conditions

$$D_0 = 0, \quad D_n = 0. \tag{2.11}$$

From (2.10) we obtain

$$\frac{j}{n} D_j + \frac{n-j}{n} D_j - \frac{j}{n} D_{j-1} = 1 + \frac{n-j}{n} D_{j+1},$$

so that

$$\frac{j}{n}(D_j - D_{j-1}) = 1 + \frac{n-j}{n}(D_{j+1} - D_j).$$

We derived a basic equation

$$D_j - D_{j-1} = \frac{n}{j} + \frac{n-j}{j}(D_{j+1} - D_j), \qquad 1 \le j \le n-1. \tag{2.12}$$

Using complete induction we prove that

$$D_1 - D_0 = \binom{n-1}{j}(D_{j+1} - D_j) + \sum_{i=1}^{j} \binom{n}{i}. \tag{2.13}$$

If $j = 1$, then the formula holds. Assume that (2.13) holds for a $j \le n-2$. We prove that it also holds for $j+1$. We use (2.12) with $j+1$ instead of j, and we have

$$\begin{aligned} D_1 - D_0 &= \binom{n-1}{j}(D_{j+1} - D_j) + \sum_{i=1}^{j} \binom{n}{i} \\ &= \binom{n-1}{j}\left[\frac{n}{j+1} + \frac{n-j-1}{j+1}(D_{j+2} - D_{j+1})\right] + \sum_{i=1}^{j} \binom{n}{i} \\ &= \binom{n-1}{j+1}(D_{j+2} - D_{j+1}) + \sum_{i=1}^{j+1} \binom{n}{i}. \end{aligned}$$

From symmetry it is clear that

$$D_1 = D_{n-1}. \tag{2.14}$$

Let $j = n-1$ in formula (2.13). Using (2.11) and (2.14), we get

$$D_1 = 2^{n-1} - 1. \tag{2.15}$$

Values D_j for $j \geq 2$ can be calculated recurrently. For example, if $n = 10$, then $D_1 = D_9 = 511$, $D_2 = D_8 = 566\frac{2}{3}$, $D_3 = D_7 = 579\frac{1}{3}$, $D_4 = D_6 = 583\frac{1}{3}$, $D_5 = 584\frac{1}{3}$. The average calculated from these expectations is $\bar{D} = (D_1 + \cdots + D_9)/9 = \frac{5065}{9} = 562.78$. To compare it with the classical random walk with $a = 10$, $z = 5$, and $p = 0.5$, formula (2.8) gives only $D_5 = 25$.

It can be surprising that D_i do not depend on i very much. The reason is that transition probabilities depend heavily on the point where the particle is located.

The name of this walk— *"reluctant"* (see *Amer. Math. Monthly* **96**, 1989, p. 162–163, Problem E3213) is connected with the fact that the random walk hesitates to stop. The more the particle approaches an absorbing barrier, the lower probability that its next movement is in the direction to this barrier.

We derive another result. Multiply formula (2.12) by j and add over j. We obtain

$$\sum_{j=1}^{n-1} j(D_j - D_{j-1}) = (n-1)n + \sum_{j=1}^{n-1} (n-j)(D_{j+1} - D_j).$$

We rewrite this as

$$nD_{n-1} - \sum_{j=0}^{n-1} D_j = (n-1)n - nD_1 + \sum_{j=1}^{n} D_j.$$

Inserting from (2.11), (2.14), and (2.15), we have

$$\bar{D} = \frac{1}{n-1}\sum_{j=1}^{n-1} D_j = \frac{n}{n-1}\left(2^{n-1} - \frac{n+1}{2}\right).$$

If $n = 10$ then $\bar{D} = \frac{5065}{9} = 562.78$. We derived this result above, but it was necessary to calculate all the values D_i.

Let q_j be the probability that A will be ruined if he has j cards. Obviously, the boundary conditions are $q_0 = 1$, $q_n = 0$ and we have the equation

$$q_j = \frac{j}{n} q_{j-1} + \frac{n-j}{n} q_{j+1}, \qquad 1 \leq j \leq n-1.$$

Similarly as in the derivation of (2.12), we obtain

$$q_{j+1} - q_j = \frac{j}{n-j}(q_j - q_{j-1}), \qquad 1 \leq j \leq n-1.$$

Define

$$\Delta_j = q_j - q_{j-1}, \qquad 1 \leq j \leq n.$$

Then we have

$$\Delta_j = \frac{(j-1)(j-2)\cdots 1}{(n-j+1)(n-j+2)\cdots(n-1)}\Delta_1, \qquad 2 \leq j \leq n,$$

Table 2.2 Probabilities q_j of ruining gambler A.

j	Δ_j	q_j	j	Δ_j	q_j
1	−0.432	0.568	6	−0.003	0.497
2	−0.048	0.521	7	−0.005	0.491
3	−0.012	0.509	8	−0.012	0.479
4	−0.005	0.503	9	−0.048	0.432
5	−0.003	0.500	10	−0.432	0.000

and so

$$\Delta_j = \frac{1}{\binom{n-1}{j-1}}\Delta_1, \qquad 1 \leq j \leq n.$$

We can see that $\Delta_n = \Delta_1$. Using formula

$$\Delta_j = \frac{n-j}{j}\Delta_{j+1}, \qquad 1 \leq j \leq n-1,$$

repeatedly, we get $\Delta_j = \Delta_{n+1-j}$, $1 \leq j \leq n$. This implies that $q_j = 1 - q_{n-j}$, $0 \leq j \leq n$. The boundary conditions lead to

$$\Delta_1 + \Delta_2 + \cdots + \Delta_n = -1$$

and so we get

$$\Delta_1 \sum_{j=1}^{n} \frac{1}{\binom{n-1}{j-1}} = -1.$$

From here Δ_1 can be calculated. We remark that

$$\sum_{k=0}^{m} \frac{1}{\binom{m}{k}} = \frac{m+1}{2^{m+1}}\left(2 + \frac{2^2}{2} + \frac{2^3}{3} + \cdots + \frac{2^{m+1}}{m+1}\right)$$

[see Kaucký (1975, p. 65), and papers cited therin]. A formula for calculating Δ_j from Δ_1 has been already derived. Since

$$\begin{aligned} \Delta_1 &= q_1 - 1, \\ \Delta_1 + \cdots + \Delta_j &= q_j - 1, \qquad 2 \leq j \leq n, \end{aligned}$$

we also have the probabilities q_j that gambler A will be ruined. For $n = 10$, we get $\sum_{j=1}^{n-1} \left[1 \Big/ \binom{n-1}{j-1}\right] = \frac{146}{63}$. Other results are introduced in Table 2.2.

2.4 RANDOM WALK UNTIL NO SHOES ARE AVAILABLE

Mr. A lives in a house with one front door and one back door. He placed n pairs of walking shoes at each door. If he wants to go for a walk, he chooses one door at random and puts on a pair of shoes. Returning back, he again chooses a door at random and leaves his shoes at the door. Sooner or later he discovers that no shoes are available at the door that he has chosen for a further walk. Find the average number of finished walks. This problem was published in *Am. Math. Monthly* (**94**, 1987, pp. 78–79), under Problem E3043.

We solve a slightly more general problem. Let Mr. A put Y pairs of shoes at the front door and Z pairs of shoes at the back door. Let $Y + Z = 2n$. Define $X = \min(Y, Z)$. Standard arguments lead to the following system of equations:

$$\begin{aligned} D_0 &= 0 \times \frac{1}{2} + \left(1 + \frac{1}{2}D_0 + \frac{1}{2}D_1\right) \times \frac{1}{2}, \\ D_j &= 1 + \frac{1}{4}D_{j-1} + \frac{1}{2}D_j + \frac{1}{4}D_{j+1} \qquad \text{for } 1 \le j \le n-1, \\ D_n &= 1 + \frac{1}{2}D_{n-1} + \frac{1}{2}D_n. \end{aligned}$$

We rewrite the equations in the form

$$\begin{aligned} D_1 - D_0 &= 2D_0 - 2, \\ D_{j+1} - D_j &= D_j - D_{j-1} - 4 \qquad \text{for } 1 \le j \le n-1, \\ 0 &= D_n - D_{n-1} - 2. \end{aligned}$$

Adding them, we obtain

$$D_0 = 2n.$$

Then the first equation of the system gives

$$D_1 = 6n - 2.$$

The next equations are

$$D_j - 2D_{j-1} + D_{j-2} = -4 \qquad \text{for } 2 \le j \le n.$$

This is nonhomogeneous linear difference equation. We can check that $D_j = -2j^2$ is a particular solution. The characteristic equation has a twofold root equal to 1, and so its general solution is

$$D_j = a + bj - 2j^2 \qquad \text{for } 2 \le j \le n.$$

Constants a and b will be determined so that the formula holds also for D_0 and D_1. This leads to $a = 2n$, $b = 4n$ and thus

$$D_j = 2n + 4nj - 2j^2 \qquad \text{for } 2 \le j \le n.$$

The answer to the original question is obtained if $j = n$. The expected number of walks is

$$D_n = 2n(n+1).$$

For instance, for $n = 5$ we have $D_5 = 60$.

Consider a variant of this problem. Mr. A walks barefooted if no shoes are available. In this case it does not matter which door he chooses when he returns. Let p_k be the probability that A's kth walk is barefooted. This is proved in theory of Markov chains (after Andrei Markov, 1856–1922) that $\lim_{k\to\infty} p_k$ exists and that $\lim_{k\to\infty} p_k = 1/(2n+1)$. See also *Am. Math. Monthly* (**94**, 1987, pp. 78–79).

2.5 THREE-TOWER PROBLEM

The gambler's ruin problem involved two gamblers. It can be generalized (not uniquely) to a larger number of gamblers. In this section we describe results published in Engel (1993).

The original problem was formulated in 1970 in the following way. There are three piles with a, b, and c chips, respectively. Each second a pile X is selected at random, another pile Y is selected at random and a chip is moved from X to Y. Find the expected waiting time $f(a, b, c)$ until one pile is empty. The name of this three-tower problem, originated from this formulation.

We can also consider three gamblers A, B, and C. At the beginning, A, B, and C have a, b, and c coins, respectively. Let $a > 0$, $b > 0$, and $c > 0$.

Case 1. A randomly chosen gambler loses a coin and another randomly chosen gambler gets it. The game continues until one of the three gamblers is ruined. Find the expected length of the game $f(a, b, c)$.

The state of the game at the beginning is given by the point (a, b, c). After one step the game is in one of the neighboring points

$$\begin{array}{lll} (a, b+1, c-1), & (a, b-1, c+1), & (a+1, b, c-1), \\ (a-1, b, c+1), & (a+1, b-1, c), & (a-1, b+1, c). \end{array}$$

Transition to any of these six points has the same probability 1/6. The set of these six points will be denoted by $S(a, b, c)$. So the rules of the game lead to the equation

$$f(a, b, c) = 1 + \frac{1}{6} \sum_{S(a,b,c)} f(x, y, z) \tag{2.16}$$

with boundary conditions

$$f(a, b, 0) = f(a, 0, c) = f(0, b, c) = 0. \tag{2.17}$$

We prove that a solution of equation (2.16) under boundary conditions (2.17) is

$$f(a, b, c) = \frac{3abc}{a+b+c} \tag{2.18}$$

and that this solution is unique.

Inserting (2.18) into equation (2.16), it is easy to verify that it is a solution. It is clear that the conditions (2.17) are fulfilled. It remains only to prove that the solution is unique. Assume that $f_1(a,b,c)$ is another solution. Define

$$h(a,b,c) = f(a,b,c) - f_1(a,b,c).$$

It is obvious that

$$h(a,b,c) = \frac{1}{6} \sum_{S(a,b,c)} h(x,y,z). \tag{2.19}$$

If the starting point (a,b,c) is given, then the function h is defined only on a finite set of points. So at a point (a_0,b_0,c_0) it reaches its maximum, which we denote by M. In view of (2.19), the function h also reaches the value M at every point belonging to $S(a_0,b_0,c_0)$. An analogous conclusion can be reached for each point from $S(a_0,b_0,c_0)$, and we can continue until we reach the boundary, where at least one of three components is zero. However, h is equal to 0 on the boundary. So we proved that $h(x,y,z) \leq 0$ holds for all points (x,y,z) from the set where h is defined. The inequality $h(x,y,z) \geq 0$ can be proved similarly. Then $f(x,y,z) = f_1(x,y,z)$ holds everywhere and the solution is unique.

Solution (2.18) was not derived from theory, but rather guessed from simulations on a computer. We saw that it was not difficult to show that it was a solution and that the solution was unique. However, it was possible to guess the result by the method used by Read (1966). We describe his procedure in Section 2.6.

Case 2. At the beginning, gamblers play in the same way as in case 1. If a gambler is ruined, the remaining two gamblers continue to play in the classical way. The game continues until a gambler wins everything and has all $a+b+c$ coins. Let $g(a,b,c)$ be the expected length of the play. The function g must satisfy the equation

$$g(a,b,c) = 1 + \frac{1}{6} \sum_{S(a,b,c)} g(x,y,z), \tag{2.20}$$

but now with boundary conditions

$$g(a,b,0) = ab, \qquad g(a,0,c) = ac, \qquad g(0,b,c) = bc. \tag{2.21}$$

The boundary conditions (2.21) follow from the fact that after a gambler is ruined we have the classical gambler's ruin problem, the solution of which is given by formula (2.8). From simulations it was guessed that the solution is

$$g(a,b,c) = ab + bc + ca. \tag{2.22}$$

It can be easily verified that (2.22) is a solution and that it is unique. It was more difficult to guess solution (2.22) than (2.18) because such a simple form was not expected.

Case 3. Each player bets a coin and has the same probability $\frac{1}{3}$ to win pool with three coins. If one gambler is ruined, the remaining two continue to play in the classical way until one of them has all $a+b+c$ coins.

Here we move from the point (a, b, c) into a point from the set

$$T(a,b,c) = \{(a+2, b-1, c-1),\ (a-1, b+2, c-1),\ (a-1, b-1, c+2)\}.$$

Transition to each of the three points is equally probable. If $k(a, b, c)$ is the expected length of the play, we have

$$k(a,b,c) = 1 + \frac{1}{3} \sum_{T(a,b,c)} k(x,y,z)$$

with boundary conditions

$$k(a,b,0) = ab, \qquad k(a,0,c) = ac, \qquad k(0,b,c) = bc.$$

The play was simulated on computer and it was conjectured that

$$k(a,b,c) = ab + bc + ca - \frac{2abc}{a+b+c-2}. \tag{2.23}$$

In the standard way it can be proved that this is a unique solution.

Case 4. At the beginning, the gamblers behave as in the case 3. However, they play until one of them is ruined. In this case the expected length l of the play satisfies the equation

$$l(a,b,c) = 1 + \frac{1}{3} \sum_{T(a,b,c)} l(x,y,z)$$

with boundary conditions

$$l(a,b,0) = l(a,0,c) = l(0,b,c) = 0.$$

It was found that

$$l(a,b,c) = \frac{abc}{a+b+c-2}.$$

This result was communicated to A. Engel by a former International Mathematical Olympiad contestant, M. Stoll. Maybe he used methods similar to those introduced in Section 2.6.

It is not difficult to generalize formula (2.2) to the situation with k players for $k \geq 4$.

A *four-tower problem* analogous to case 1 seems to remain unsolved. The expected length of the play must be a complicated function because it was calculated that $f(3,2,2,2) = 350,612/69,969$. No simple formula can give such a complicated result for so small values of a, b, c, d. A. Engel (1993) was able to find only an approximation to the expected length.

2.6 GAMBLER'S RUIN PROBLEM WITH TIES

Consider two gamblers, A and B. Assume that A has x coins and B has y coins. Let $x > 0, y > 0$ and $a = x + y$. Each gambler tosses a coin. If one of them has a head and the other a tail, head wins and the winning gambler gets both coins. If the coins are alike, neither gambler wins and we have *tie*. Let D_x be the expected number of tosses of two coins until a gambler is ruined. As in Section 2.1, we obtain equation

$$D_x = 1 + \frac{1}{4}D_{x-1} + \frac{1}{2}D_x + \frac{1}{4}D_{x+1}$$

with boundary conditions $D_0 = 0$, $D_a = 0$. This equation can be rewritten in the form

$$D_{x+1} - 2D_x + D_{x-1} = -4, \qquad D_0 = 0, D_a = 0.$$

It is easy to verify that a particular solution of this difference equation is $D_x = -2x^2$, and so its general solution is $D_x = Ax + B - 2x^2$. From the boundary conditions we determine constants A and B. This gives

$$D_x = 2x(a - x) = 2xy.$$

Consider three gamblers A, B, and C with x, y, and z coins, respectively. Each gambler tosses a coin. Whenever all coins appear alike, the gamblers repeat the throw, otherwise the odd person wins. Find the expected number of tosses $D(x, y, z)$ until a gambler is ruined. It is clear that there is probability $\frac{1}{4}$ of a tie. Probability that A wins is $\frac{1}{4}$, and the same holds also for B and C. So we have

$$\begin{aligned} D(x,y,z) &= 1 + \frac{1}{4}[D(x,y,z) + D(x+2, y-1, z-1) \\ &\quad + D(x-1, y+2, z-1) + D(x-1, y-1, z+2)] \end{aligned}$$

with boundary conditions

$$D(0, y, z) = 0, \quad D(x, 0, z) = 0, \quad D(x, y, 0) = 0.$$

To solve the equation for $D(x, y, z)$, Read (1966) proposes the following trick. The problem with two gamblers was solved by $D_x = 2xy$. Now look for the solution in the form $D(x, y, z) = Kxyz$ where K is a constant. Inserting into equation for $D(x, y, z)$, we obtain

$$K = \frac{4}{3(x + y + z - 2)} = \frac{4}{3(a - 2)}.$$

This means that a solution is

$$D(x, y, z) = \frac{4xyz}{3(x + y + z - 2)}.$$

We can see that the boundary conditions are also fulfilled. Note that the solution is similar to case 4 in Section 2.5. Read (1966) points out that this method cannot be used for four gamblers having x, y, z, and u coins with analogous rules. It is easy to verify that $D(x, y, z, u)$ does not have the form $Kxyzu$ where K is a constant.

Table 2.3 Results of games.

aaa	baa	bba
aab	bab	bbb
aba		
abb		

2.7 PROBLEM OF PRIZE DIVISION

This book started with the *problem of prize division* and so we only briefly review the problem here. Players A and B play a series of games. There are no draws. The play is fair. The games are independent. The player who is the first to win six games is the winner and gets a prize. The play was interrupted in the moment when A won five games and B won three games. The problem is how to divide the prize fairly. The play could have three additional games maximum. If a and b denote that A and B win a game, respectively, then all possibilities introduced in Table 2.3 are equally probable.

Player B gets the prize only if bbb is realized. Then the probability of this event is $\frac{1}{8}$ and fair division of the prize after the result 5:3 is 7:1 for player A.

We find the expected number of games needed to reach the final decision from state 5:3. The possibilities where the winner of the prize is determined immediately after the first, second, and third games are introduced in the first, second, and third columns of Table 2.3, respectively. The expected number of additional games is

$$1 \cdot \frac{4}{8} + 2 \cdot \frac{2}{8} + 3 \cdot \frac{2}{8} = 1.75.$$

A natural question is to find the expected number of all games when players start from the state 0:0. We solve this problem in a more general case.

Players A and B play a series of games. Let p be the probability that A wins a game, and let $q = 1 - p$ be probability that B wins a game. Assume that game results are independent of one another. A wins a prize if he wins r games before B wins s games. In the opposite case B wins the prize. Consider only the case that $r \geq 1$, $s \geq 1$. We find probability that A wins the prize and the expected number of games needed to finish the play.

The state of the play can be represented by a lattice point in the rectangle with sides r and s. Point (i, j) corresponds to situation when A won i games and B won j games. The case with $r = 8$ and $s = 6$ is demonstrated in Fig. 2.1. If $i \leq r - 1$ and $j \leq s - 1$, then we can move from point (i, j) either to point $(i + 1, j)$ with probability p or to point $(i, j+1)$ with probability q. Other transitions are impossible. The play continues until the right or the upper side of the rectangle is reached. Points on the right side (denoted by empty circles) indicate that A is the winner of the prize. Points on the upper side (black circles) indicate that B is the winner of the prize. The play cannot reach point (r, s).

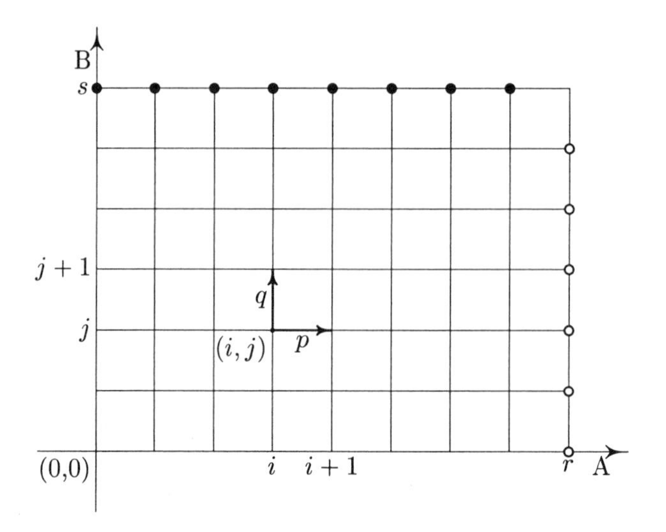

Fig. 2.1 Division of a prize.

The play starts at the point (0,0). We introduce random variables

$$U_{ij} = \begin{cases} 1 & \text{if the play visits point } (i,j), \\ 0 & \text{otherwise.} \end{cases}$$

Rules of the play give

$$\begin{aligned}
\mathsf{P}(U_{ij} = 1) &= \binom{i+j}{i} p^i q^j, \quad i = 0, \ldots, r-1;\ j = 0, \ldots, s-1; \\
\mathsf{P}(U_{rj} = 1) &= \mathsf{P}(U_{r-1,j} = 1)p = \binom{r-1+j}{r-1} p^r q^j, \quad j = 0, \ldots, s-1; \\
\mathsf{P}(U_{is} = 1) &= \mathsf{P}(U_{i,s-1} = 1)q = \binom{i+s-1}{i} p^i q^s, \quad i = 0, \ldots, r-1, \\
\mathsf{P}(U_{rs} = 1) &= 0.
\end{aligned}$$

The length of play (meaning the number of games) is

$$U = \sum_{i=0}^{r-1} \sum_{j=0}^{s-1} U_{ij}.$$

Another equivalent expression is

$$U = \sum_{i=0}^{r} \sum_{j=0}^{s} U_{ij} - 1.$$

The expected length of the play is

$$\mathsf{E}U = \sum_{i=0}^{r-1} \sum_{j=0}^{s-1} \mathsf{E}U_{ij} = \sum_{i=0}^{r-1} \sum_{j=0}^{s-1} \mathsf{P}(U_{ij} = 1).$$

Probabilities $\mathsf{P}(A)$ and $\mathsf{P}(B)$ that A and B get the prize, respectively, are

$$\begin{aligned}\mathsf{P}(A) &= \sum_{j=0}^{s-1}\mathsf{P}(U_{rj}=1) = p^r\sum_{j=0}^{s-1}\binom{r-1+j}{r-1}q^j,\\ \mathsf{P}(B) &= \sum_{i=0}^{r-1}\mathsf{P}(U_{is}=1) = q^s\sum_{i=0}^{r-1}\binom{i+s-1}{i}p^i.\end{aligned}$$

Assume that $r = s = n$. We derive another formula for the length of the play and prove that expected number of games is maximal in the case that $p = 0.5$. Although this assertion is intuitively clear, its proof is not very simple. We need two auxiliary theorems.

Theorem 2.1 *Let* $n \geq 1$. *Define* $S_1 = 1$ *and*

$$S_n = \sum_{k=0}^{n-1}\binom{n+k}{n}p^k q^k(p^{n-k}+q^{n-k}), \qquad n \geq 2.$$

If $n \geq 2$, *then*

$$S_n = S_{n-1} + \frac{1}{n}\binom{2n-2}{n-1}p^{n-1}q^{n-1}. \tag{2.24}$$

Proof. If $n = 2$ then (2.24) can be verified directly. Assume that $n \geq 3$. For arbitrary j, we have

$$p^j + q^j = p^{j-1} + q^{j-1} - pq(p^{j-2}+q^{j-2}).$$

So we get

$$\begin{aligned}S_n &= \sum_{k=0}^{n-2}\binom{n+k}{n}p^k q^k(p^{n-k}+q^{n-k}) + \binom{2n-1}{n}p^{n-1}q^{n-1}(p+q)\\ &= \sum_{k=0}^{n-2}\binom{n+k}{n}p^k q^k(p^{n-k-1}+q^{n-k-1})\\ &\quad -\sum_{k=0}^{n-2}\binom{n+k}{n}p^{k+1}q^{k+1}(p^{n-k-2}+q^{n-k-2}) + \binom{2n-1}{n}p^{n-1}q^{n-1}\\ &= \binom{n}{n}(p^{n-1}+q^{n-1}) + \sum_{k=1}^{n-2}\binom{n+k}{n}p^k q^k(p^{n-k-1}+q^{n-k-1})\\ &\quad -2\binom{2n-2}{n}p^{n-1}q^{n-1}\\ &\quad -\sum_{k=0}^{n-3}\binom{n+k}{n}p^{k+1}q^{k+1}(p^{n-k-2}+q^{n-k-2}) + \binom{2n-1}{n}p^{n-1}q^{n-1}\end{aligned}$$

which can be written as

$$\begin{aligned} S_n &= \binom{n}{n}(p^{n-1}+q^{n-1}) + \sum_{k=1}^{n-2}\left[\binom{n+k}{n} - \binom{n+k-1}{n}\right] p^k q^k (p^{n-k-1} \\ &\quad + q^{n-k-1}) + \left[\binom{2n-1}{n} - 2\binom{2n-2}{n}\right] p^{n-1}q^{n-1} \\ &= \sum_{k=0}^{n-2}\binom{n+k-1}{n-1} p^k q^k (p^{n-1-k} + q^{n-1-k}) + \frac{1}{n}\binom{2n-2}{n-1} p^{n-1}q^{n-1} \\ &= S_{n-1} + \frac{1}{n}\binom{2n-2}{n-1} p^{n-1}q^{n-1}. \quad \square \end{aligned}$$

Theorem 2.2 *If $n \geq 1$, then*

$$\sum_{k=0}^{n-1}\binom{n+k}{n}(p^n q^k + q^n p^k) = \sum_{k=0}^{n-1}\binom{2k}{k}\frac{p^k q^k}{k+1}.$$

Proof. Our formula is a direct consequence of Theorem 2.1. □

Theorem 2.3 *Assume that $r = s = n$. Let U be the number of games of the play. Then we have for all $k = 0, 1, \ldots, n-1$ that*

$$\begin{aligned} \mathsf{P}(U = n+k) &= \binom{n-1+k}{k}(p^n q^k + p^k q^n), \\ \mathsf{E}U &= n\sum_{k=0}^{n-1}\binom{n+k}{n}(p^n q^k + p^k q^n). \end{aligned}$$

The value $\mathsf{E}U$ is maximal if $p = 0.5$.

Proof. We start with

$$\begin{aligned} \mathsf{P}(U = n+k) &= \mathsf{P}(U_{nk} = 1) + \mathsf{P}(U_{kn} = 1) \\ &= \mathsf{P}(U_{n-1,k} = 1)p + \mathsf{P}(U_{k,n-1} = 1)q \\ &= \binom{n-1+k}{k} p^{n-1} q^k p + \binom{n-1+k}{k} p^k q^{n-1} q \\ &= \binom{n-1+k}{k}(p^n q^k + p^k q^n). \end{aligned}$$

Further, we obtain

$$\mathsf{E}U = \sum_{k=0}^{n-1}(n+k)\binom{n-1+k}{k}(p^n q^k + p^k q^n) = n\sum_{k=0}^{n-1}\binom{n+k}{n}(p^n q^k + p^k q^n).$$

Since $\mathsf{E}U = nS_n$, Theorem 2.2 yields that $\mathsf{E}U$ is maximal if $p = 0.5$. □

We used the method published in *Am. Math. Monthly* (**99**, 1992, pp. 272–274, Problem E3386).

We introduce a small numerical illustration. For $r = 1$, $s = 3$, and $p = 0.5$ we get $\mathsf{E}U = 1.75$, which is known from our starting considerations. For $r = s = 6$ and $p = 0.5$, we obtain $\mathsf{E}U = 9.293$. This is the expected number of games from the beginning of the play. For $r = s = 6$ and $p = 0.75$, we have $\mathsf{E}U = 7.907$. For $r = s = 12$ and $p = 0.5$, we get $\mathsf{E}U = 20.132$, and $r = s = 12$ with $p = 0.75$ gives $\mathsf{E}U = 15.985$.

The problem of prize division can be generalized in the following way (see Mačák 1997, pp. 108–109). Players $A_1, \ldots, A_n$ play a series of games. The player who wins k games first wins a prize. Assume that player A_i wins a game with probability p_i $(i = 1, \ldots, n)$ and that game results are independent of one another. The play was interrupted at the moment when player A_i must win a_i games to get the prize $(i = 1, \ldots, n)$. We find probability P_1 that A_1 would get the prize under these circumstances. She gets the prize in the case that $a_1 + u_2 + \cdots + u_n$ games will be played in which she wins a_1 games (but one of them must be the last game), and player A_i wins u_i games $(0 \leq u_i \leq a_i - 1;\ i = 2, \ldots, n)$. This means that in $a_1 - 1 + u_2 + \cdots + u_n$ games played before the last one, player A_1 wins $a_1 - 1$ games, player A_2 wins u_2 games, and so on, and player A_n wins u_n games. The corresponding probability is given by the *multinomial distribution*

$$\frac{(a_1 + u_2 + \cdots + u_n - 1)!}{(a_1 - 1)!\, u_2! \ldots u_n!} p_1^{a_1 - 1} p_2^{u_2} \ldots p_n^{u_n}.$$

At the same time player A_1 must win the last game, which has probability p_1. Different vectors $(u_2, \ldots, u_n)$ represent disjoint events, and so we get

$$P_1 = \sum_{i=2}^{n} \sum_{u_i=0}^{a_i - 1} \frac{(a_1 + u_2 + \cdots + u_n - 1)!}{(a_1 - 1)!\, u_2! \ldots u_n!} p_1^{a_1} p_2^{u_2} \ldots p_n^{u_n}.$$

2.8 TENNIS

It is widely known that standard scores in a game of *tennis* are 0, 15, 30, and 40. The deuce score is 40:40. In such a situation a player must win twice consecutively to have game. If he wins once, he has advantage. If he loses the next ball, we will have a deuce again.

Instead of standard scores, we use points 0,1,2,..., which is more common in mathematics. Player A wins a game with score 4:0, 4:1, 4:2, 5:3, 6:4, 7:5, and so on. Player B wins a game with score 0:4,1:4, 2:4, 3:5, 4:6, 5:7, and so forth. The game is illustrated in graph form in Fig. 2.2.

Assume that A wins a point with probability p and B wins a point with probability $q = 1 - p$. Further assume that the results are independent of one another and that the probability of winning a point depends neither on the history of the game nor on the given score. Let $P(i, j)$ be the probability that A wins the game given the score

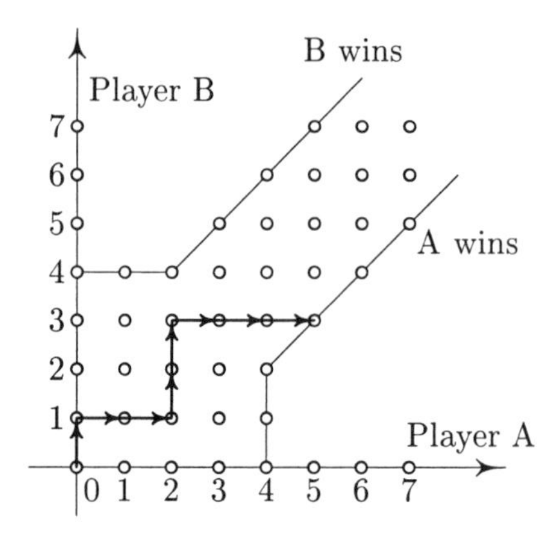

Fig. 2.2 Tennis.

$i{:}j$. If $0 \le i \le 3, 0 \le j \le 3$; or if $|i-j| \le 1$, we have obviously

$$P(i,j) = pP(i+1,j) + qP(i,j+1).$$

By a twofold application of this formula we obtain

$$\begin{aligned} P(3,3) &= pP(4,3) + qP(3,4) \\ &= p^2P(5,3) + 2pqP(4,4) + q^2P(3,5). \end{aligned}$$

Since $P(5,3) = 1$, $P(3,5) = 0$, $P(4,4) = P(3,3)$, we have $P(3,3) = p^2 + 2pqP(3,3)$, so that

$$P(3,3) = \frac{p^2}{1-2pq} = \frac{p^2}{p^2+q^2}.$$

We are especially interested in probability $P(0,0)$. We calculate it in the following way. Let $Q(i,j)$ be the probability that the game visits point (i,j). It is easy to calculate the following values:

$$\begin{array}{lll} Q(4,0) = p^4, & Q(4,1) = 4p^4q, & Q(4,2) = 10p^4q^2, \\ Q(0,4) = q^4, & Q(1,4) = 4pq^4, & Q(2,4) = 10p^2q^4, \\ Q(3,3) = 20p^3q^3. & & \end{array}$$

Player A wins a game if the graph visits one of the points (4,0), (4,1), (4,2) or if she wins after the first deuce. This leads to the formula

$$\begin{aligned} P(0,0) &= Q(4,0) + Q(4,1) + Q(4,2) + P(3,3)Q(3,3) \\ &= p^4(1 + 4q + 10q^2) + \frac{20p^5q^3}{p^2+q^2}. \end{aligned}$$

If we change p and q in this formula, we get the probability that B wins the game. The result is $1 - P(0,0)$.

Table 2.4 Probability of winning a game.

n	1	2	3	4
$P(0,0)$	$\frac{1}{2}$	$\frac{208}{243}$	$\frac{243}{256}$	$\frac{51{,}968}{53{,}125}$

Table 2.5 Probability of winning a game.

p	$P(0,0)$	p	$P(0,0)$	p	$P(0,0)$
0.1	0.001	0.4	0.264	0.7	0.901
0.2	0.022	0.5	0.500	0.8	0.978
0.3	0.099	0.6	0.736	0.9	0.999

Some time ago it was popular to express probability $P(0,0)$ in terms of the parameter $n = p/q$. Since $p + q = 1$, we have $p = n/(n+1)$, $q = 1/(n+1)$. Inserting these values into formula for $P(0,0)$ we obtain

$$\begin{aligned} P(0,0) &= \frac{n^7 + 5n^6 + 11n^5 + 15n^4}{n^7 + 5n^6 + 11n^5 + 15n^4 + 15n^3 + 11n^2 + 5n + 1} \\ &= \frac{n^4(n^3 + 5n^2 + 11n + 15)}{(1+n)^5(1+n^2)}. \end{aligned}$$

The formula for $P(0,0)$ was derived by James Bernoulli (1654–1705). His main work is *Ars Conjectandi*, whereas his *Letter to a Friend on the Game of Tennis* is much less known. He proved the theorem that the relative frequency of an event in a sequence of independent trials converges in probability to the probability of the event. He is also known for the Bernoulli numbers. To simplify his derivation of the formula for $P(0,0)$, he assumed that n is an integer. We give some values of $P(0,0)$ in Table 2.4. This table is introduced in statistical papers and books probably for historical reasons. However, we are rather interested in the direct dependence of $P(0,0)$ on p. This relationship is introduced in Table 2.5.

Now, we calculate the expected length of the game. Let $D(i,j)$ be the expected length of the game from the moment when score was $i{:}j$. If $0 \le i \le 3$, $0 \le j \le 3$ or if $|i - j| \le 1$, we get

$$D(i,j) = 1 + pD(i+1,j) + qD(i,j+1).$$

Then

$$\begin{aligned} D(3,3) &= 1 + pD(4,3) + qD(3,4) \\ &= 1 + p[1 + pD(5,3) + qD(4,4)] + q[1 + pD(4,4) + qD(3,5)]. \end{aligned}$$

Since

$$D(5,3) = D(3,5) = 0, \quad D(3,3) = D(4,4),$$

Table 2.6 Expected length of the game.

p	$D(0,0)$	p	$D(0,0)$	p	$D(0,0)$
0.1	4.46	0.4	6.48	0.7	5.83
0.2	5.09	0.5	6.75	0.8	5.09
0.3	5.83	0.6	6.48	0.9	4.46

we have

$$D(3,3) = 1 + p + q + 2pqD(3,3).$$

It can be proved that $D(3,3)$ is a finite number. Then

$$D(3,3) = \frac{2}{1-2pq} = \frac{2}{p^2+q^2}.$$

It is not difficult to verify that $D(3,3)$ is maximal in the case that $p = q = \frac{1}{2}$. Here we have $D(3,3) = 4$.

We would like to know $D(0,0)$. This expectation is

$$\begin{aligned} D(0,0) &= 4[Q(4,0) + Q(0,4)] + 5[Q(4,1) + Q(1,4)] \\ &\quad +6[Q(4,2) + Q(2,4)] + [6 + D(3,3)]Q(3,3). \end{aligned}$$

Inserting terms in this formula yields

$$D(0,0) = 4(p^4+q^4)+20(p^4q+pq^4)+60(p^4q^2+p^2q^4)+40p^3q^3\left(3+\frac{1}{p^2+q^2}\right).$$

Values $D(0,0)$ that are dependent on p are introduced in Table 2.6.

Our presentation is based on a paper by Croucher (1985) and on Section 4.7 in the book by Blom et al. (1994). The authors of the book cited the book by Hald (1990, p. 241).

2.9 WOLF AND SHEEP

Let $n \geq 4$ equidistant points be marked on a circle. Case $n = 4$ is exhibited in Fig. 2.3. A black piece representing a *wolf*, is laid at one of the points, say at point 1. White pieces, representing *sheep*, are laid at remaining points. The wolf takes a random walk on the circle. This means that it goes in each step with probability $\frac{1}{2}$ to one neighboring point and with probability $\frac{1}{2}$ to the other neighboring point. If this is its first visit at this point, there is a sheep there and the wolf eats the poor sheep. Which of the points should a clever lamb occupy to ensure the largest probability that the wolf visits it last?

We introduce a solution given by V. Beneš. The case $n = 4$, which is demonstrated in Fig. 2.3, will be considered first. Let p_k be the probability that the wolf visits the

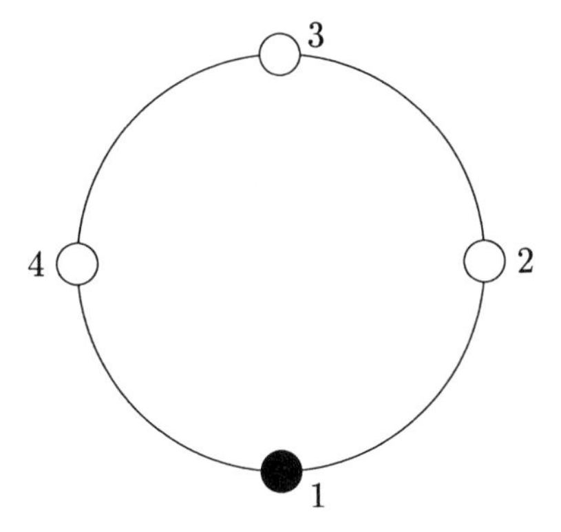

Fig. 2.3 Starting position.

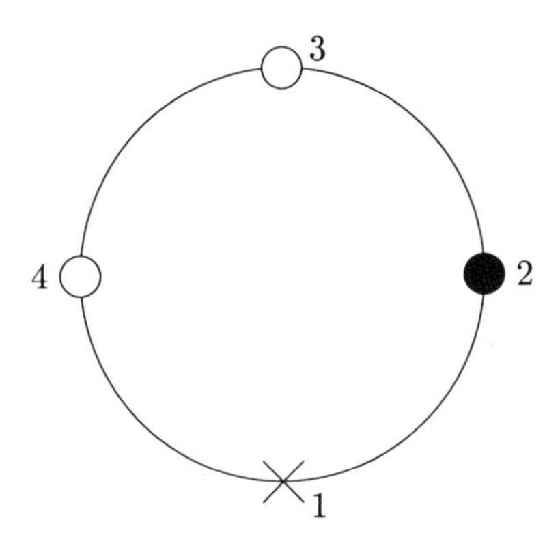

Fig. 2.4 Position after one step.

point k last $(k = 2, 3, 4)$. It can be proved that the wolf visits all points in finite many steps with probability 1 and thus

$$p_2 + p_3 + p_4 = 1. \tag{2.25}$$

If the wolf goes to point 2 in the first step, we get the scenario situation in Fig. 2.4.

Since the wolf goes to either point 2 or point 4 in the first step, the theorem of total probability yields

$$p_3 = \frac{1}{2}p_2 + \frac{1}{2}p_4. \tag{2.26}$$

From the problem's symmetry, we can see that

$$p_2 = p_4. \tag{2.27}$$

Solving the system (2.25), (2.26), (2.27) we get the surprising result that

$$p_2 = p_3 = p_4 = \frac{1}{3}. \tag{2.28}$$

This means that the wolf visits any of the points $2, 3, 4$ last with the same probability. Our reasoning leading to the result (2.28) was correct but not complete. If we wanted to calculate probability p_4 analogously, we would seemingly get it from the theorem of total probability

$$p_4 = \frac{1}{2} \times 0 + \frac{1}{2} \times p_3, \tag{2.29}$$

since if the wolf visits point 4 in the first step, this point is visited last with probability 0. The system (2.25), (2.26), (2.27), (2.28) has no solution. Why is this consideration false? Assume again that the wolf visits point 2 in the first step as shown in Fig. 2.4. If this is taken as the starting position, then the wolf comes to the sheep standing at point 4 last not only if it visits point 4 last but also if it visits point 4 in the penultimate step and only after it comes to point 1. That's why formula (2.29) is not valid. Such a situation cannot arise when we consider a point that is not a neighbor of point 1. Therefore, formula (2.26) is correct and the result (2.28) is also correct.

In the general case when the wolf starts at point 1 and $n \geq 4$, we obtain

$$1 = p_2 + p_3 + \cdots + p_n, \tag{2.30}$$

$$p_k = p_{n+2-k}, \qquad k = 2, \ldots, n, \tag{2.31}$$

$$p_k = \frac{1}{2} p_{k-1} + \frac{1}{2} p_{k+1}, \qquad k = 3, 4, \ldots, n-2. \tag{2.32}$$

The general solution of the homogeneous equation (2.32) is

$$p_k = A + Bk,$$

since the characteristic equation $\lambda^2 - 2\lambda + 1$ has one double root $\lambda_{12} = 1$. From condition (2.31), when $k = 2$, we get $B = 0$, so that $p_k = A$. Formula (2.30) then implies $A = 1/(n-1)$. Thus the solution is

$$p_2 = p_3 = \cdots = p_n = \frac{1}{n-1}.$$

Even a clever sheep cannot improve its chances with these calculations, since each point designated for the sheep has the same probability that the wolf will visit it last.

3

Principle of reflection

3.1 TICKET-SELLING AUTOMAT

Consider a ticket-dispensing automat and that each ticket costs one crown. The automat is technically constructed in such a way that a customer is allowed to use either a one-crown or a two-crown coin. If a customer pays with a two-crown coin, the automat returns a ticket and a one-crown coin from its store. If it has no one-crown coins in its store, it returns a two-crown coin and becomes blocked.

To illustrate the problem, we suppose that the automat will be used by 30 people and that we know that 15 of them have one-crown coins and 15 have only two-crown coins. Let the order of customers be completely random. If we want to guarantee that the automat will not become blocked, we must ensure that it has at least 15 one-crown coins in its store at the beginning (of operation). If it has less than 15 coins and the first 15 customers come with two-crown coins, mechanical blocking would result.

In practical situations it is nearly impossible to ensure perfect reliability. We try only to ensure that reliability is sufficiently high. Instead of 100% reliability, we want to ensure 99% reliability or 95% reliability. We ask following question: How many one-crown coins in the automat's store are sufficient for 95% probability to ensure that the automat will not become blocked? And how many one-crown coins are needed for 99% probability?

You can stop reading for a while and try to guess the result. It is not easy because few people have experience with the assumption of random order of the queue.

In this chapter we solve the problem, not only for this case when the stucture of the queue is known in advance, but also for situations when we know only the probability that a customer has one-crown coin. This is so-called queue with random structure.

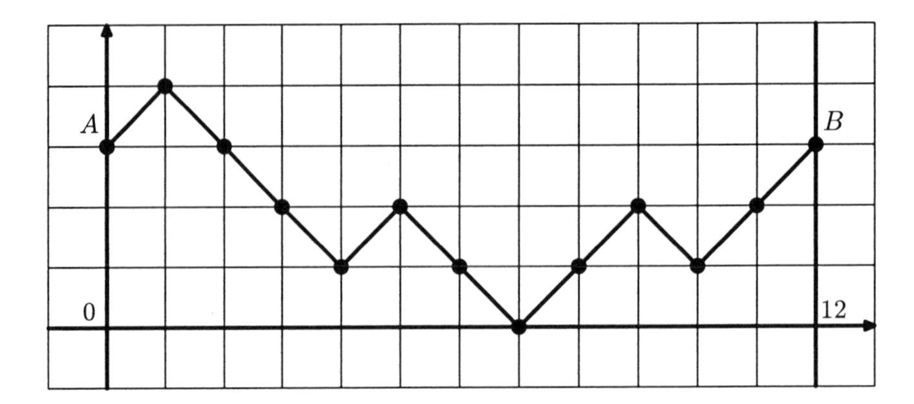

Fig. 3.1 Trajectory of a queue.

Finally, we will even assume that the length of the queue with random structure may be random.

3.2 KNOWN STRUCTURE OF THE QUEUE

We solve a general formulation of the problem introduced above. Assume that the automat has a one-crown coins in its store at the beginning and that it should serve a queue with N customers. Let m customers have one-crown coins so that the number of customers with two-crown coins is $n = N - m$.

The number of one-crown coins in the automat's store changes by one. If a customer with a one-crown coin uses the automat, the reserve in the store increases by one coin. After service for a customer with a two-crown coin the store reserve decreases by one coin. This process can be illustrated graphically: an example is given in Fig. 3.1. The store starts with $a = 3$ one-crown coins, and we have a queue with $N = 12$ customers. Here $m = 6$ customers had one-crown coins and $n = 6$ customers had two-crown coins. From the figure we can infer that a customer with a one-crown coin used the authomat first, then three customers with two-crown coins, and so on. After service for the seventh customer, the automat's store was empty. However, the automat was not blocked, since the next customers supplied one-crown coins.

Each order in the queue is represented by a *trajectory*, which starts at point $A = (0, a)$, then goes up m times and down n times, so that it terminates at point $B = (N, a + m - n)$. Some trajectories can fall under the horizontal x axis. They correspond to orderings where the automat is blocked. If the order in the queue is random, all trajectories have the same probability. The number of trajectories is the same as the number of ways that m rises can be distributed at N places, and it gives $\binom{N}{m}$.

If $a + m - n < 0$, we have nothing to solve because the automat will surely become blocked. The number of customers with two-crown coins is larger than the number of customers with one crown-coins together with the original contents of the automat's store. So we consider only cases where $a + m - n \geq 0$. If we insert $n = N - m$, we have situations where $m \geq (N - a)/2$.

First we determine the number of trajectories that fall under the horizontal axis. Each of them must touch or intersect the line z, which is parallel with the horizontal axis and is located one unit below it, (see Fig. 3.2). As soon as the trajectory touches the line z for the first time, we watch—besides its continuation—also its graph symmetric with respect to z. This auxiliary graph terminates at point

$$C = (N, -a - m + n - 2),$$

which is symmetric to the point $B = (N, a + m - n)$ with respect to the line z. It is clear that there exists exactly one trajectory from point A to point C corresponding to a trajectory from A to B that falls at least once under the horizontal axis. This result is known as the *principle of reflection*. Lozansky and Rousseau (1996, p. 146), write that the author of this principle is Antoine Désiré André. If a trajectory from A to C has u segments up and d segments down, then

$$\begin{aligned} u + d &= N && \text{(trajectory has } N \text{ segments total)}, \\ a + u - d &= -a - m + n - 2 && \text{(trajectory terminates at point } C\text{)}. \end{aligned}$$

The solution of this system is

$$u = n - a - 1, \qquad d = m + a + 1.$$

This implies that the number of trajectories from A to C is $\binom{N}{m+a+1}$. Probability $Q_a(N, m)$, that the automat will become blocked, is

$$Q_a(N, m) = \frac{\dbinom{N}{m + a + 1}}{\dbinom{N}{m}} = \frac{n(n - 1) \dots (n - a)}{(m + 1)(m + 2) \dots (m + a + 1)}.$$

We are interested in probability $P_a(N, m)$ that the automat will not become blocked. This result is given by

$$P_a(N, m) = 1 - Q_a(N, m),$$

so that

$$P_a(N, m) = 1 - \frac{n(n - 1) \dots (n - a)}{(m + 1)(m + 2) \dots (m + a + 1)}. \tag{3.1}$$

We have already mentioned that this formula holds only for $a + m - n \geq 0$, that is, for $m \geq (N - a)/2$. If $m < (N - a)/2$, then, of course, $P_a(N, m) = 0$.

For given N and m, we can insert values $a = 0, 1, 2, \dots, N - m$ and find when the probability $P_a(N, m)$ is equal to or larger than α, say, $\alpha = 0.95$ or $\alpha = 0.99$ for the first time. Some selected results are given in Tables 3.1 and 3.2.

Looking into Table 3.1 we can answer the question asked in the first section of this chapter. If the length of the queue is $N = 30$ and there are $m = 15$ customers in it with

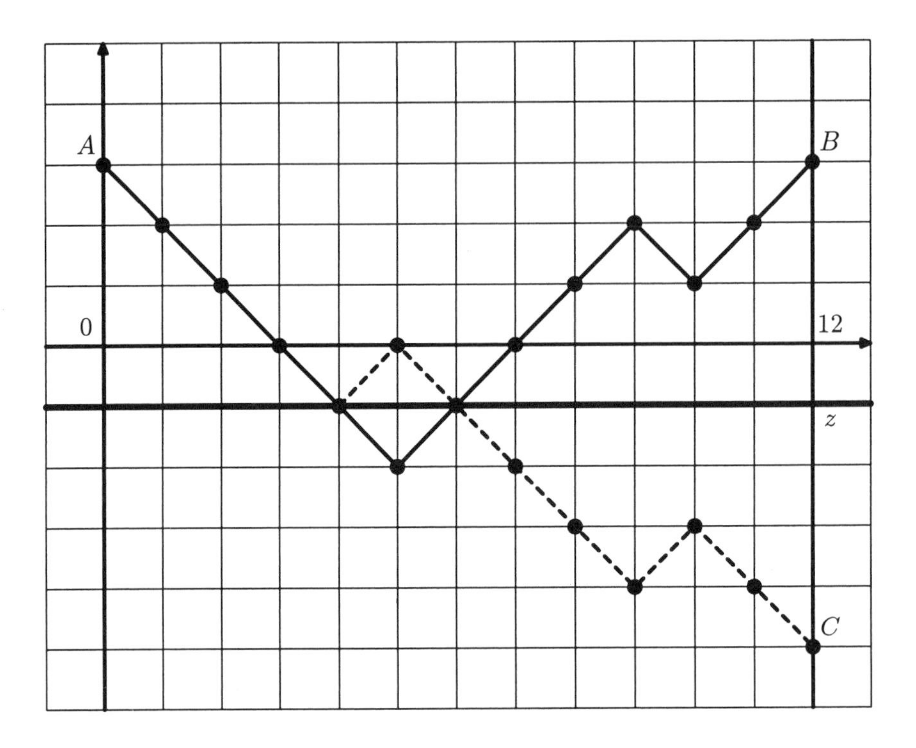

Fig. 3.2 Principle of reflection.

Table 3.1 Starting supply of one-crown coins for $\alpha \geq 0.95$.

Number of customers N	m	Minimal value a, ensuring probability 0.95	Actual probability $P_a(N, m)$
30	15	6	0.962
60	30	9	0.965
120	60	13	0.962
30	10	13	0.980
60	20	23	0.964
120	40	43	0.953

Table 3.2 Starting supply of one-crown coins for $\alpha \geq 0.99$.

Number of customers N	m	Minimal value a, ensuring probability 0.99	Actual probability $P_a(N, m)$
30	15	8	0.996
60	30	11	0.992
120	60	16	0.992
30	10	14	0.995
60	20	25	0.996
120	40	45	0.992

one-crown coins, only 6 one-crown coins in the automat's store at the beginning will suffice to guarantee service without blocking with probability 0.962. People usually guess that such a probability can be guaranteed only when a is near 15 (which ensures probability 1). A surprisingly low value ($a = 6$) is a consequence of the assumption that the order in the queue is random. In such cases, extremely unfavorable orders of two-crown coins in the queue will have only very small probability.

This problem has many modifications. One of them is itroduced in the journal *Teaching Statist.* (**13**, 1991, p. 63, Problem PPP2), called "Food for thought".

> William Brown is charging his enemies 50p for a Sunday afternoon view of his "Man Eating Snail". He has an initial float of two 50p coins (bribes from Ethel and Robert to leave the house); and, during the morning, he will sucker eleven separate advance bookers, five of whom have just 50p coins, and six only £1 pieces. Each time that William cannot give change, he sends Ginger to the corner shop with the £1 bit to buy 50p's worth of gob-stoppers.

Inserting into our formulas, we infer that Ginger never gets to buy sweets with probability $P_2(11, 2) = 0.643$.

Another application of the principle of reflection is introduced in Section 8.10.

3.3 QUEUE WITH RANDOM STRUCTURE

It is rarely known in advance how many customers in a queue will have one-crown coins and two-crown coins. From certain evidence it is rather possible to estimate the probability p that a randomly chosen customer has a one-crown coin. Then $q = 1 - p$ is the probability that a customer has a two-crown coin.

We can ask again how many one-crown coins the automat must have at the beginning in its store in order to serve all N customers in the queue without becoming blocked with a probability of at least α. If we demand $\alpha = 1$, automat would have to contain N one-crown coins at the beginning. So we analyze only the case $\alpha \in (0, 1)$.

Table 3.3 Starting supply of one-crown coins for $\alpha \geq 0.95$.

Number of customers N	Probability p	Minimal value a, ensuring probability 0.95	Actual probability $P_{a,p}(N)$
30	$\frac{1}{2}$	10	0.957
60	$\frac{1}{2}$	15	0.960
120	$\frac{1}{2}$	21	0.955
30	$\frac{1}{3}$	18	0.952
60	$\frac{1}{3}$	32	0.956
120	$\frac{1}{3}$	58	0.962

Our calculation is based on theorem of total probability (see Theorem 1.2). We use also notation introduced there.

In our situation B_m is the event that in a queue with N customers there are precisely m of them with one-crown coins. Event A is that automat will not be blocked. We know probability $\mathsf{P}(A|B_m)$ because it was calculated in the previous section and denoted by $P_a(N, m)$. If $m \geq (N-a)/2$, then we calculate it from formula (3.1), otherwise it is zero.

Under our assumptions the number m of customers with one-crown coins in a queue with N customers is a random variable with binomial distribution. This means that we can write

$$\mathsf{P}(B_m) = \binom{N}{m} p^m q^{N-m}, \qquad m = 0, 1, \ldots, N.$$

The probability that an automat having a one-crown coins at the beginning will not become blocked is

$$P_{a,p}(N) = \sum_{m=\lceil \frac{N-a}{2} \rceil}^{N} \left[1 - \frac{(N-m)(N-m-1)\ldots(N-m-a)}{(m+1)(m+2)\ldots(m+a+1)}\right] \binom{N}{m} p^m q^{N-m}, \tag{3.2}$$

where $\lceil y \rceil$ denotes the smallest integer that is larger than or equal to y. If N is not too large, we can calculate the probability $P_{a,p}(N)$ from formula (3.2).

We are interested in such values a that ensure high probability of service without interruption. For some selected numbers N and p, the results are introduced in Tables 3.3 and 3.4.

Table 3.4 Starting supply of one-crown coins for $\alpha \geq 0.99$.

Number of customers N	Probability p	Minimal value a, ensuring probability 0.99	Actual probability $P_{a,p}(N)$
30	$\frac{1}{2}$	14	0.995
60	$\frac{1}{2}$	19	0.990
120	$\frac{1}{2}$	28	0.992
30	$\frac{1}{3}$	22	0.995
60	$\frac{1}{3}$	37	0.992
120	$\frac{1}{3}$	64	0.992

Compare Tables 3.3 and 3.4 with Tables 3.1 and 3.2, respectively. We can see that the random structure of the queue leads to the need for a larger supply of one-crown coins at the beginning in order to keep the same reliability of service. Larger uncertainty in structure of the queue causes larger variability which must be compensated by a larger supply at the beginning.

3.4 RANDOM NUMBER OF CUSTOMERS

Now, we consider a slightly more complicated case. Assume that the number N of customers is not known in advance. Let N customers occur with probability f_N, and let customers join the queue independently. We assume again that a customer has a one-crown coin with probability p and a two-crown coin with probability $q = 1 - p$. Define $f = (f_0, f_1, \ldots)$. Find the probability that the automat will not become blocked.

Once more using the theorem of total probability, we get

$$P_{a,p,f} = \sum_{N=0}^{\infty} P_{a,p}(N) f_N.$$

In some cases it is possible to assume that the number of customers is a random variable having Poisson distribution with parameter λ. Then

$$f_N = \frac{\lambda^N}{N!} e^{-\lambda}, \qquad N = 0, 1, \ldots.$$

The expected number of customers in such a queue is exactly equal to λ. Here it is impossible to guarantee perfect service because the automat would have to contain

Table 3.5 Starting supply of one-crown coins for $\alpha \geq 0.95$.

Parameter λ	Probability p	Minimal value a, ensuring probability 0.95	Actual probability $P_{a,p,\lambda}$
30	$\frac{1}{2}$	10	0.954
60	$\frac{1}{2}$	15	0.961
30	$\frac{1}{3}$	20	0.966
60	$\frac{1}{3}$	33	0.952

Table 3.6 Starting supply of one-crown coins for $\alpha \geq 0.99$.

Parameter λ	Probability p	Minimal value a, ensuring probability 0.99	Actual probability $P_{a,p,\lambda}$
30	$\frac{1}{2}$	14	0.993
60	$\frac{1}{2}$	20	0.993
30	$\frac{1}{3}$	23	0.991
60	$\frac{1}{3}$	39	0.992

an infinite number of one-crown coins at the beginning. The probability that automat will not become blocked is

$$P_{a,p,\lambda} = \sum_{N=0}^{\infty} P_{a,p}(N) \frac{\lambda^N}{N!} e^{-\lambda}, \tag{3.3}$$

where $P_{a,p}(N)$ is as given by formula (3.2). Some numerical results can be found in Tables 3.5 and 3.6.

It is surprising that values a introduced in Tables 3.5 and 3.6 are either equal to or not much larger than those in Tables 3.3 and 3.4. The random number of customers having Poisson distribution only slightly influences the quality of service in comparison with the corresponding queue with random structure with fixed number $N = \lambda$ of customers.

We must point out that numerical calculations based on formulas (3.1) to (3.3) fail if N or λ is too large. Then it would be necessary to derive new formulas based on asymptotic behavior of individual members. We do not discuss them in further detail here.

This chapter is based on a paper by Anděl (1990/91).

4

Records

4.1 RECORDS, PROBABILITY, AND STATISTICS

Extremal situations have always been watched with great interest — not only watched, but also recorded. In old manuscripts we read about extraordinarily cold summers with snow, about floods, high temperatures and so on. Material that is systematically compiled is particularly important for science. For example, meteorological series have been recorded in Prague (Klementinum) daily since 1775.

Of course, people are interested in records in other disciplines as well. The *Guinness Book of Records* is popular reading throughout the world. New editions appear very frequently, because there are new records and new disciplines. It is not so well known that the *Guinness Book of Records* is connected with mathematical statistics. Mr. William Gosset (1876–1937), who is known mostly under the pseudonym Student, was hired by an ancestor of Mr. Guinness as a brewer in Guinness' brewery in Dublin and later even as head brewer in Guinness' brewery in London. Student distribution, Student t test and so on can be found in all textbooks of mathematical statistics, and they are basic statistical tools in applications.

Consider a series of real numbers $X_1, \ldots, X_n$. The first value X_1 is called *record*. This is logical, because it is the largest value among all values that have been recorded until that moment. Further, we say that X_i is a record if it is the largest value among all numbers that have been recorded until time i. More exactly, X_i is a record, if

$$X_i > \max(X_1, \ldots, X_{i-1}) \quad \text{for} \quad i \geq 2.$$

Such records are sometimes called *"upper" records*. We also have *lower records*; the value X_1 is always a lower record. The value X_i $(i \geq 2)$ is a lower record if

$X_i < \min(X_1, \ldots, X_{i-1})$. Since lower records in series $X_1, \ldots, X_n$ are obviously upper records in series $-X_1, \ldots, -X_n$, we will not deal with lower records and will use the term record to mean upper record throughout.

If the series $X_1, \ldots, X_n$ is random, we can ask a lot of questions. Find the probability that there are exactly r records in the series. Find the expected number of records. How long shall we wait for the next record? We answer these questions in the following sections. To be able to derive some laws, we make some assumptions about variables $X_1, \ldots, X_n$.

1. Let $X_1, \ldots, X_n$ be independent identically distributed (i.i.d.) random variables.

2. Let the distribution of X_i be continuous.

The first assumption practically excludes records in athletic disciplines where people systematically try to improve their performance. Models used for such situations have a different character and are rather special (e.g., see Henningsen 1984). We do not discuss such problems here.

The second assumption, concerning continuous distribution introduced in the second point is only technical. It excludes the possibility that two values could be equal and that a record could be tied. Readers interested in such situations can consult Vervaat (1973).

Lectures about records have generated much interest and so records is a frequent topic in papers for popularization of mathematical statistics (see Anděl 1988b, Boas 1981, Glick 1978).

4.2 EXPECTED NUMBER OF RECORDS

Let $X_1, \ldots, X_n$ be a series of i.i.d. random variables with a continuous distribution. First we find the probability that the value X_n will be a record. All variables $X_1, \ldots, X_n$ are mutually different with probability 1. They can be ordered in $n!$ ways, and all orderings have the same probability. If X_n is a record, then it is the largest value of all. The remaining $n-1$ variables before X_n can be ordered in $(n-1)!$ ways that again have the same probability. Then the probability p_n that X_n will be a record is

$$p_n = \frac{(n-1)!}{n!} = \frac{1}{n}. \tag{4.1}$$

To simplify discussion, we introduce auxiliary variables $Y_1, \ldots, Y_n$ such that

$$Y_i = \begin{cases} 1, & \text{if } X_i \text{ is record,} \\ 0, & \text{if } X_i \text{ is not record,} \end{cases} \tag{4.2}$$

$i = 1, \ldots, n$. Variables Y_i are called *indicators*. It is easy to see that

$$\mathsf{E}Y_i = 1 \times p_i + 0 \times (1 - p_i) = p_i.$$

Table 4.1 Values n such that $S_n \geq N$ for the first time.

N	2	3	4	5	6	7	8	9	10
n	4	11	31	83	227	616	1674	4550	12367

The total number of records R_n in the series $X_1, \ldots, X_n$ is

$$R_n = Y_1 + \cdots + Y_n,$$

so that

$$\mathsf{E}R_n = \mathsf{E}Y_1 + \cdots + \mathsf{E}Y_n = 1 + \frac{1}{2} + \cdots + \frac{1}{n}. \tag{4.3}$$

The number $\mathsf{E}R_n$ is interpreted as follows. If we had a very large number of mutually independent series of length n and we found a number of records in each of them, then the average of these numbers would approach the number $\mathsf{E}R_n$ according to the *law of large numbers.*

The *harmonic series* is the series $1 + \frac{1}{2} + \frac{1}{3} + \cdots$ and it has infinite sum. This means that the number $\mathsf{E}R_n$ exceeds any given bound as n tends to infinity. It may be interesting that the computer does not indicate this result. If we tried to sum harmonic series on a computer that uses positive numbers from 10^{-99} to 10^{99}, we could consider as the last term of the series expression $1/n$ for $n = 10^{99}$. If the computer adds 10^6 terms in a second, then it would come to this last term after 10^{93} seconds, which is 3.17×10^{85} years. If we did not take rounding off into consideration, the computer would yield in this incredibly long time only the result

$$1 + \frac{1}{2} + \frac{1}{3} + \cdots + \frac{1}{10^{99}} = 228.533\ldots.$$

This number is not very large and does not indicate that the sum of the whole series is infinite. In fact, if we calculate in double-precision arithmetic and use common statistical packages, then a test for $1 \stackrel{?}{=} 1 + 1/2^i$ gives "true" for $i = 53$ and $2^{53} \doteq 9 \times 10^{15}$. A test for zero is different, and we get "true" for the first time for $0 \stackrel{?}{=} 1/2^{1024}$. Of course, some packages working with symbolic computation are able to promptly reply that the sum is infinite without any numerical calculations.

In Table 4.1, for given values N, we find n such that the sum

$$S_n = 1 + \frac{1}{2} + \frac{1}{3} + \cdots + \frac{1}{n}$$

for the first time is equal to or larger than N. Again, we come very quickly to astronomical numbers. For example, if $N = 100$, then $n = 1.509 \times 10^{43}$.

We read in Table 4.1 that in a random series (such as maximal high of snow in winter) an 11-year-old child has seen in the life three records on average; a gentleman 31 years old, four records; and a man 83 years old, five records. Isn't this a key to the fact that in our youth winters were colder with more snow, summers were warmer, and, generally, everything was much better?

Sum S_n occurs in mathematics very frequently. Leonhard Euler (1707–1783) derived the following approximation, called the *Euler formula*

$$S_n \doteq \ln n + \gamma, \tag{4.4}$$

where $\gamma = 0.57721566\ldots$ is the *Euler constant.* If $N > 0$ is given and we are looking for n such that $S_n \doteq N$, then the Euler formula gives the approximation $n \doteq e^{N-\gamma}$.

If we wish to know how precise formula (4.4) is, we can use inequalities

$$\frac{1}{2n+\frac{2}{5}} \le 1 + \frac{1}{2} + \cdots + \frac{1}{n} - \ln n - \gamma < \frac{1}{2n+\frac{1}{3}}$$

(see *Am. Math. Monthly* **99**, 1992, p. 684, Problem E3432). It is known that the constant $\frac{1}{3}$ on the right-hand side cannot be replaced by a slightly larger number, whereas the constant $\frac{2}{5}$ on the left-hand side can be replaced by a slightly smaller number $(2\gamma - 1)/(1-\gamma) = 0.36527\ldots$ and equality holds only when $n = 1$.

Sometimes a more exact variant of the formula (4.4) is used, namely

$$S_n = \ln n + \gamma + \frac{1}{2n} + \frac{1}{12n^2} + O(n^{-4})$$

(see *Am. Math. Monthly* **100**, 1993, p. 298, Problem E3448).

4.3 PROBABILITY OF r RECORDS

The expected number of records introduced in formula (4.3) does not indicate the probability $p_{r,n}$ that in a series $X_1, \ldots, X_n$ there will be exactly r records. We now describe how this probability can be computed.

The probability that only one record occurs is

$$p_{1,n} = \mathsf{P}(R_n = 1) = \frac{1}{n}.$$

We can derive this result similarly as that at the beginning of Section 4.2. One record indicates that X_1 is the largest value. Variables $X_1, \ldots, X_n$ can be ordered in $n!$ ways. If the largest value is the first, then the remaining variables can be ordered in $(n-1)!$ ways. Then probability $p_{1,n}$ is $(n-1)!/n! = 1/n$.

It is also easy to calculate the probability that there will be n records in a random series with n terms. Variables $X_1, \ldots, X_n$ can be ordered in $n!$ ways, but only one of them is such that it gives n records. This is the case when variables are ordered increasingly from the smallest to the largest one. Therefore

$$p_{n,n} = \mathsf{P}(R_n = n) = \frac{1}{n!}.$$

The remaining probabilities must be calculated using a more complicated method.

Table 4.2 Probabilities $p_{r,n}$.

n	r 1	2	3	4	5
1	1	—	—	—	—
2	0.5	0.5	—	—	—
3	0.33333	0.5	0.166 67	—	—
4	0.25	0.45833	0.25	0.04167	—
5	0.2	0.41667	0.29167	0.08333	0.00833

Theorem 4.1 *If n is arbitrary positive integer and if r is arbitrary positive integer such that $r \leq n$, then*

$$p_{r,n} = \frac{n-1}{n} p_{r,n-1} + \frac{1}{n} p_{r-1,n-1}, \tag{4.5}$$

where

$$p_{1,1} = 1, \qquad p_{r,0} = 0. \tag{4.6}$$

Proof. Let $A_{i,j}$ be the event that there will be exactly i records in the series $X_1, \ldots, X_j$. Let B_n be the event that variable X_n will be a record; B_n^c is the event that X_n will not be a record. We have

$$p_{r,n} = \mathsf{P}(R_n = r) = \mathsf{P}(A_{r,n-1} \cap B_n^c) + \mathsf{P}(A_{r-1,n-1} \cap B_n).$$

We know that

$$\mathsf{P}(A_{r,n-1} \cap B_n^c) = \mathsf{P}(A_{r,n-1} | B_n^c)\mathsf{P}(B_n^c).$$

The probability that there will be r records in series $X_1, \ldots, X_{n-1}$ is not influenced by the fact that variable X_n is not a record. Then

$$\mathsf{P}(A_{r,n-1} | B_n^c) = \mathsf{P}(A_{r,n-1}) = p_{r,n-1}.$$

Formula (4.1) says that $\mathsf{P}(B_n) = 1/n$. It gives $\mathsf{P}(B_n^c) = 1 - \mathsf{P}(B_n) = (n-1)/n$, and we have

$$\mathsf{P}(A_{r,n-1} \cap B_n^c) = \frac{n-1}{n} p_{r,n-1}.$$

Analogously

$$\begin{aligned} \mathsf{P}(A_{r-1,n-1} \cap B_n) &= \mathsf{P}(A_{r-1,n-1} | B_n)\mathsf{P}(B_n) \\ &= \mathsf{P}(A_{r-1,n-1})\mathsf{P}(B_n) = \frac{1}{n} p_{r-1,n-1}. \quad \square \end{aligned}$$

We can calculate probabilities $p_{r,n}$ from (4.5) when n is not very large number. Some numerical results are introduced in Tables 4.2 and 4.3. Since the relation $\mathsf{E}R_n = \sum_{r=1}^{n} r p_{r,n}$ holds, we can check numerically that it agrees with formerly derived formula $\mathsf{E}R_n = S_n$.

If n is large, the recurrent calculation of probabilities $p_{r,n}$ is not appropriate. Time of calculation can be rather long, and some inaccuracies caused by rounding off can appear. We discuss asymptotic behaviour of $p_{r,n}$ in the next section.

Table 4.3 Probabilities $p_{r,n}$ for $n = 10, 100, 200, 500$.

r	$p_{r,10}$	$p_{r,100}$	$p_{r,200}$	$p_{r,500}$
1	0.10000	0.01000	0.00500	0.00200
2	0.28290	0.05177	0.02937	0.01358
3	0.32316	0.12585	0.08213	0.04447
4	0.19943	0.19299	0.14674	0.09403
5	0.07422	0.21120	0.18925	0.14491
6	0.01744	0.17672	0.18859	0.17408
7	0.00260	0.11815	0.15168	0.17017
8	0.00024	0.06510	0.10151	0.13948
9	0.00001	0.03024	0.05781	0.09799
10	0.00000	0.01206	0.02851	0.06000
11	—	0.00418	0.01234	0.03250
12	—	0.00128	0.00474	0.01572
13	—	0.00035	0.00163	0.00686
14	—	0.00008	0.00051	0.00272
15	—	0.00002	0.00014	0.00099
16	—	0.00000	0.00004	0.00033
17	—	0.00000	0.00000	0.00010
18	—	0.00000	0.00000	0.00003
19	—	0.00000	0.00000	0.00001

4.4 STIRLING NUMBERS

The expression $x^{(n)} = x(x-1)\dots(x-n+1)$ is called nth downward power and expression $x^{[n]} = x(x+1)\dots(x+n-1)$ is called nth upward power. Define

$$\Delta x^{(n)} = (x+1)^{(n)} - x^{(n)}.$$

It is easy to check that

$$\Delta x^{(n)} = n x^{(n-1)}. \tag{4.7}$$

Analogously, we have $\Delta x^{(n+1)} = (n+1)x^{(n)}$, and so

$$x^{(n)} = \frac{\Delta x^{(n+1)}}{n+1}. \tag{4.8}$$

Formula (4.8) implies that

$$\sum_{x=0}^{M-1} x^{(n)} = \frac{M^{(n+1)}}{n+1}, \qquad n \geq 1. \tag{4.9}$$

Formula (4.9) can also be proved by complete induction.

Consider upward powers. From the formula

$$0 + 1 + 2 + \cdots + M = \frac{M(M+1)}{2},$$

which can be also rewritten in the form

$$0^{[1]} + 1^{[1]} + 2^{[1]} + \cdots + M^{[1]} = \frac{M^{[2]}}{2},$$

we can guess that

$$\sum_{x=0}^{M} x^{[n]} = \frac{M^{[n+1]}}{n+1}, \qquad n \geq 1. \tag{4.10}$$

This formula can be proved by complete induction. If $M = 1$, then the left-hand side is

$$0 + 1^{[n]} = 1 \times 2 \times \cdots \times n = n!$$

and the right-hand side is

$$\frac{1^{[n+1]}}{n+1} = \frac{(n+1)!}{n+1} = n!.$$

Assume that (4.10) holds for a positive integer M. Then

$$\sum_{x=0}^{M+1} x^{[n]} = \sum_{x=0}^{M} x^{[n]} + (M+1)^{[n]} = \frac{M^{[n+1]}}{n+1} + (M+1)^{[n]} = \frac{(M+1)^{[n+1]}}{n+1}.$$

Formula (4.7) is an analog of the derivative of usual power x^n and formulas (4.9) and (4.10) are analogs of integration of function x^n.

Upward and downward powers are useful in probability theory for calculating *factorial moments* $\mathsf{E}X^{(n)}$, which can be obtained in some cases (e.g., in binomial and Poisson distributions) more easily than the usual moments. However, it is necessary to express $x^{(n)}$ as the usual polynomial and to write x^n as linear combination of downward powers $x^{(i)}$. Here we consider only the first problem. We express $x^{(n)}$ in the form

$$x(x-1)(x-2)\cdots(x-n+1) = S_n^n x^n + S_n^{n-1} x^{n-1} + \cdots + S_n^1 x. \tag{4.11}$$

Coefficients S_n^r are called the *Stirling* (after James Stirling, 1692–1770) *numbers of the first kind*. Multiplying, we get

$$S_1^1 = 1, \qquad S_2^2 = 1,\ S_2^1 = -1, \qquad S_3^3 = 1,\ S_3^2 = -3,\ S_3^1 = 2.$$

Formula (4.11) yields

$$S_n^r = S_{n-1}^{r-1} - (n-1) S_{n-1}^r,$$

where $S_j^i = 0$ for $i \leq 0$ and for $i > j$. Numbers S_n^r are introduced in Table 4.4.

Define $D_{r,n} = |S_n^r|$. Since $D_{r,n} = (-1)^{n+r} S_n^r$, we have $S_n^r = (-1)^{n+r} D_{r,n}$. This implies that numbers $D_{r,n}$ satisfy recurrent relation

$$D_{r,n} = (n-1) D_{r,n-1} + D_{r-1,n-1},$$

Table 4.4 Stirling numbers of the first kind S_n^r.

n	r 1	2	3
1	1		
2	-1	1	
3	2	-3	1

where $D_{1,1} = 1$ and $D_{r,0} = 0$. If we compare these formulas with (4.5), we can see that

$$p_{r,n} = \frac{D_{r,n}}{n!}. \tag{4.12}$$

This formula was derived by Rényi. Since the properties of Stirling numbers are well known, we can get an idea about properties $p_{r,n}$. For instance, from a formula derived by Jordan in 1947, we have

$$p_{r,n} \sim \frac{1}{(r-1)!\,n}(\ln n + \gamma)^{r-1}$$

as $n \to \infty$. Even for large values n, this approximation gives sufficiently accurate results only for small numbers r. For example, if $n = 500$, then, comparing with entries in Table 4.3 we can see that agreement to five decimal places is found only for $r = 1$ and for $r = 2$. If $r = 3$, approximation gives 0.04613 instead of the correct 0.04447 and quality deteriorates rapidly. If necessary, more complicated procedures must be used.

4.5 INDICATORS

Indicators $Y_1, \ldots, Y_n$ were introduced by formula (4.2). Remember that $Y_i = 1$ if X_i is a record; otherwise $Y_i = 0$. We know that $\mathsf{P}(Y_i = 1) = 1/i$, $\mathsf{E}Y_i = 1/i$. An easy calculation gives

$$\mathsf{var}\, Y_i = \mathsf{E}Y_i^2 - (\mathsf{E}Y_i)^2 = \frac{1}{i} - \frac{1}{i^2}. \tag{4.13}$$

Theorem 4.2 *If $i \neq j$, then random variables Y_i and Y_j are independent.*

Proof. Let $1 \leq i < j \leq n$. We have

$$\begin{aligned}
&\mathsf{P}(Y_i = 1, Y_j = 1) \\
&\quad= \mathsf{P}\{X_i = \max(X_1, \ldots, X_i), X_j = \max(X_1, \ldots, X_j)\} \\
&\quad= \mathsf{P}\{X_i = \max(X_1, \ldots, X_i) < X_j = \max(X_{i+1}, \ldots X_j)\}.
\end{aligned}$$

Table 4.5 Expectations and variances of the number of records.

n	$\mathsf{E}R_n$	$\mathsf{var}\,R_n$	n	$\mathsf{E}R_n$	$\mathsf{var}\,R_n$	n	$\mathsf{E}R_n$	$\mathsf{var}\,R_n$
2	1.50	0.25	20	3.60	2.00	200	5.88	4.24
3	1.83	0.47	30	3.99	2.38	300	6.28	4.64
4	2.08	0.66	40	4.28	2.66	400	6.57	4.93
5	2.28	0.82	50	4.50	2.87	500	6.79	5.15
6	2.45	0.96	60	4.68	3.05	600	6.97	5.33
7	2.59	1.08	70	4.83	3.20	700	7.13	5.49
8	2.72	1.19	80	4.97	3.33	800	7.26	5.62
9	2.83	1.29	90	5.08	3.45	900	7.38	5.74
10	2.93	1.38	100	5.19	3.55	1 000	7.49	5.84

Thus we have

$$\begin{aligned}
&\mathsf{P}(Y_i = 1, Y_j = 1)\\
&\quad = \mathsf{P}\{X_i = \max(X_1, \dots, X_i)\}\\
&\qquad \times \mathsf{P}\{\max(X_1, \dots, X_i) < \max(X_{i+1}, \dots, X_j)\}\\
&\qquad \times \mathsf{P}\{X_j = \max(X_{i+1}, \dots, X_j)\}\\
&\quad = \frac{1}{i}\frac{j-i}{j}\frac{1}{j-i} = \mathsf{P}(Y_i = 1)\mathsf{P}(Y_j = 1).
\end{aligned}$$

Since $\mathsf{P}(Y_i = 0, Y_j = 1) + \mathsf{P}(Y_i = 1, Y_j = 1) = \mathsf{P}(Y_j = 1)$, we also get

$$\begin{aligned}
\mathsf{P}(Y_i = 0, Y_j = 1) &= \mathsf{P}(Y_j = 1) - \mathsf{P}(Y_i = 1, Y_j = 1)\\
&= \mathsf{P}(Y_j = 1) - \mathsf{P}(Y_i = 1)\mathsf{P}(Y_j = 1)\\
&= \mathsf{P}(Y_j = 1)[1 - \mathsf{P}(Y_i = 1)] = \mathsf{P}(Y_i = 0)\mathsf{P}(Y_j = 1).
\end{aligned}$$

Relations $\mathsf{P}(Y_i = 1, Y_j = 0) = \mathsf{P}(Y_i = 1)\mathsf{P}(Y_j = 0)$ and $\mathsf{P}(Y_i = 0, Y_j = 0) = \mathsf{P}(Y_i = 0)\mathsf{P}(Y_j = 0)$ can be proved similarly. $\square$

Theorem 4.3 *The variance of the total number of records R_n is*

$$\mathsf{var}\,R_n = \sum_{i=1}^{n} \frac{1}{i} - \sum_{i=1}^{n} \frac{1}{i^2}.$$

Proof. We know that $R_n = Y_1 + \cdots + Y_n$ and that variables $Y_1, \dots, Y_n$ are pairwise independent. Then they are also uncorrelated, which implies that $\mathsf{var}\,R_n = \mathsf{var}\,Y_1 + \cdots + \mathsf{var}\,Y_n$. It suffices to insert terms from (4.13). $\square$

Expectations $\mathsf{E}R_n$ and variances $\mathsf{var}\,R_n$ for some values n are introduced in Table 4.5.

The value $\mathsf{E}R_n = S_n$ can be approximated using Euler formula (4.4). Since

$$\sum_{i=1}^{n} \frac{1}{i^2} \to \frac{\pi^2}{6} = 1.6449\dots \qquad \text{as } n \to \infty$$

for sufficiently large n, we have an approximation

$$\operatorname{var} R_n \doteq \ln n + \gamma - \frac{\pi^2}{6}.$$

Knowledge of $\mathsf{E}R_n$ and $\operatorname{var} R_n$ can be used to estimate some probabilities. For example, if $r \geq \mathsf{E}R$, then *one-sided Tshebyshev* [after Pafnuti L. Tshebyshev (also transliterated as Chebyshev), 1821–1894] *inequality* (see Feller 1968, Vol. II) yields

$$\mathsf{P}(R_n \geq r) \leq \frac{\operatorname{var} R_n}{\operatorname{var} R_n + (r - \mathsf{E}R_n)^2}.$$

For example, if we insert $n = 60$, we get $\mathsf{P}(R_{60} \geq 9) \leq 0.14$. However, this estimated probability is very conservative. Exact numerical computation gives $\mathsf{P}(R_{60} \geq 9) = 0.022$.

It can be proved that all indicators $Y_1, \dots, Y_n$ are independent variables (see Glick 1978). Since

$$\frac{\operatorname{var} Y_n}{(\mathsf{E}R_n)^2} \sim \frac{1}{n(\ln n)^2},$$

we have

$$\sum_{n=1}^{\infty} \frac{\operatorname{var} Y_n}{\left(\sum_{i=1}^{n} \mathsf{E}Y_i\right)^2} < \infty$$

and from the *Kolmogorov convergence criterion* for sums of independent variables (see Loève 1955) we obtain that $R_n/\mathsf{E}R_n \to 1$ with probability 1. Thus the number of records R_n tends to infinity with probability 1 as $n \to \infty$ in the same speed as $\ln n$.

4.6 WHEN RECORDS OCCUR

Let N_r be the index of the variable that creates the rth record. If variables $X_1, \dots, X_n$ are measured in equidistant time intervals, then N_r is simply the time when the rth record occurs. From definition we have $N_1 = 1$ since the first observation is always a record. Throughout this section we assume that the series $X_1, X_2, \dots$ is infinite.

First, we consider the problem when the second record occurs. If it occurs at time $t \geq 2$, then X_t is the largest value among $X_1, \dots, X_t$ and X_1 is the second largest. Variables between them can be ordered in $(t-2)!$ ways. There are $t!$ equally probable orderings of variables $X_1, \dots, X_t$, and so we get

$$\mathsf{P}(N_2 = t) = \frac{(t-2)!}{t!} = \frac{1}{t(t-1)}.$$

Using the formula

$$\frac{1}{t(t-1)} = \frac{1}{t-1} - \frac{1}{t},$$

it can be easily verified that

$$\sum_{t=2}^{\infty} \mathsf{P}(N_2 = t) = 1.$$

This means that the second record occurs in finite time with probability 1. It agrees with the results introduced in previous sections. However

$$\mathsf{E}N_2 = \sum_{t=2}^{\infty} t\mathsf{P}(N_2 = t) = \sum_{t=2}^{\infty} \frac{1}{t-1} = \infty.$$

A short explanation of this surprising result is based on the fact that probabilities $\mathsf{P}(N_2 = t)$ decrease very slowly when t increases, and in a series, the waiting time for the second record may be extremely long.

Theorem 4.4 *If $r \geq 2$ and $t \geq r$, we have*

$$\mathsf{P}(N_r = t) = \frac{p_{r-1,t-1}}{t},$$

where probabilities $p_{r,n}$ are introduced in Theorem 4.1.

Proof. We have

$$\begin{aligned} \mathsf{P}(N_r = t) &= \mathsf{P}(r\text{th record occurs at time } t) \\ &= \mathsf{P}(\text{variables } X_1, \ldots, X_{t-1} \text{ contain } r-1 \text{ records,} \\ &\qquad \text{variable } X_t \text{ is a record}) \\ &= \mathsf{P}(\text{variables} X_1, \ldots, X_{t-1} \text{ contain } r-1 \text{ records}) \\ &\quad \times \mathsf{P}(\text{variable } X_t \text{ is a record}) \\ &= p_{r-1,t-1}\frac{1}{t}. \quad \square \end{aligned}$$

Using (4.8), we obtain

$$\mathsf{P}(N_r = t) = \frac{D_{r-1,t-1}}{t!}.$$

Since $N_r \geq N_2$ for $r \geq 2$, we have

$$\mathsf{E}N_r \geq \mathsf{E}N_2 = \infty.$$

The expected waiting time for the rth record ($r \geq 2$) is infinite. However,

$$\sum_{t=r}^{\infty} \mathsf{P}(N_r = t) = 1.$$

Since it is impossible to know anything about average behavior of variables N_r from their expectations, it is necessary to use another characteristic. Probabilities

Table 4.6 Medians m_r of occurrence of the rth record.

r	2	3	4	5	6	7	8	9
m_r	2	7	20	57	152	424	1166	3200

$\mathsf{P}(N_r = t)$ can be computed using Theorems 4.4 and 4.1. We find integer m_r such that the sum

$$\mathsf{P}(N_r = r) + \mathsf{P}(N_r = r+1) + \cdots + \mathsf{P}(N_r = m_r)$$

for the first time is equal to or larger than $\frac{1}{2}$. The number m_r is the *median* of the distribution of the occurrence of the rth record. If we investigate a large number of independent random series, then in about half of them we observe the rth record no later than at time m_r and in about half of them no sooner than at time m_r. Table 4.6 gives an idea about numbers m_r.

We mentioned already above that records in series $X_1, X_2, \ldots$ occur less and less frequently. In some sense they can be compared with the occurrence of powers of 2 in positive integers. Such an analogy is actually valid, but is analogous with powers of number e. Rényi (1962) proved that $N_r^{1/r} \to e$ with probability 1.

We derive another formula for random variable N_2. We have

$$\begin{aligned} \mathsf{P}(N_2 > n) &= \sum_{i=n+1}^{\infty} \mathsf{P}(N_2 = i) \\ &= \sum_{i=n+1}^{\infty} \frac{1}{(i-1)i} = \sum_{i=n+1}^{\infty} \left(\frac{1}{i-1} - \frac{1}{i} \right) = \frac{1}{n}. \end{aligned} \tag{4.14}$$

This result is generalized in Section 4.8.

4.7 TEMPERATURE RECORDS IN PRAGUE

We compare theoretical results with real data. In a book entitled *Meteorological observations in Prague — Klementinum* (in Czech), we can find monthly maxima of temperature from January 1775 to December 1975. From here we derived yearly maxima of temperature, which are listed in Table 4.7 (records are boxed). The same series is represented graphically in Fig. 4.1, where records are marked by small solid circles.

In the given time interval, seven records occurred. The expected number of records in a random series with 201 terms is $\mathsf{E}R_{201} = S_{201} = 5.88$. We can say that the observed number of records agrees with this expected result. It agrees also with data given in Table 4.6. The seventh record occurs with the same probability before as after the 424th observation; if we consider the sixth record, then the corresponding median is the 152th observation.

Table 4.7 Yearly maxima of temperature in Prague — Klementinum.

	0	1	2	3	4	5	6	7	8	9
1770						31.3	32.4	29.3	31.4	29.3
1780	28.0	34.4	35.6	32.5	30.8	30.9	26.5	32.2	33.4	31.7
1790	32.2	33.0	33.3	37.0	32.5	31.1	31.0	31.9	31.5	31.8
1800	32.3	33.7	34.7	31.9	31.5	29.1	29.0	34.2	34.1	31.2
1810	30.1	33.3	29.8	27.8	29.5	29.4	27.2	28.7	29.9	33.5
1820	30.4	26.4	30.6	30.4	29.7	30.5	32.2	33.6	33.6	32.5
1830	35.7	29.6	34.8	32.0	33.1	31.6	31.1	30.3	34.4	30.9
1840	28.2	35.1	32.3	28.9	29.5	35.0	30.9	30.8	33.6	30.7
1850	30.6	28.6	31.9	35.3	30.7	32.7	30.1	32.6	30.5	33.2
1860	30.4	34.8	32.5	32.9	29.8	35.2	30.3	32.8	34.5	32.8
1870	35.1	31.2	32.8	35.0	32.6	34.3	30.8	32.4	31.0	29.8
1880	31.1	33.4	29.6	31.6	32.0	32.1	32.4	33.5	30.5	34.7
1890	31.0	32.0	35.9	30.1	32.7	31.8	32.1	30.6	31.1	31.2
1900	32.8	31.6	29.5	28.6	32.9	34.5	32.6	32.7	32.4	29.8
1910	29.1	33.5	29.0	30.5	29.6	32.3	29.7	32.2	33.0	31.6
1920	34.6	34.7	35.1	33.0	32.6	30.6	30.4	31.9	34.5	34.0
1930	34.4	32.2	33.4	34.6	33.2	37.2	33.4	33.0	32.2	31.2
1940	31.4	32.4	31.5	35.0	31.6	33.5	34.1	34.9	33.3	34.8
1950	35.7	33.1	35.6	32.7	32.4	30.5	30.0	37.6	33.5	34.4
1960	31.5	32.7	33.6	33.9	33.2	34.3	31.6	34.4	33.1	33.6
1970	32.2	33.9	33.5	31.4	34.3	30.6				

Similar results were obtained for other meteorological series. Note, however, that monthly data are influenced by periodicity of 12 months. For this reason, series must be considered for each month separately.

4.8 HOW LONG WE WAIT FOR THE NEXT RECORD

We know from formula (4.14) that we must wait for the second record at least n units of time with probability $1/n$. Generally, *interrecord waiting time* is defined as

$$W_r = N_{r+1} - N_r.$$

Variable W_r represents the time elapsed between the rth and $(r+1)$th records.

Theorem 4.5 *Conditional distribution of the variable W_r given that the rth record occurred at time N_r is*

$$\mathsf{P}(W_r = k|N_r) = \frac{N_r}{(N_r + k - 1)(N_r + k)}.$$

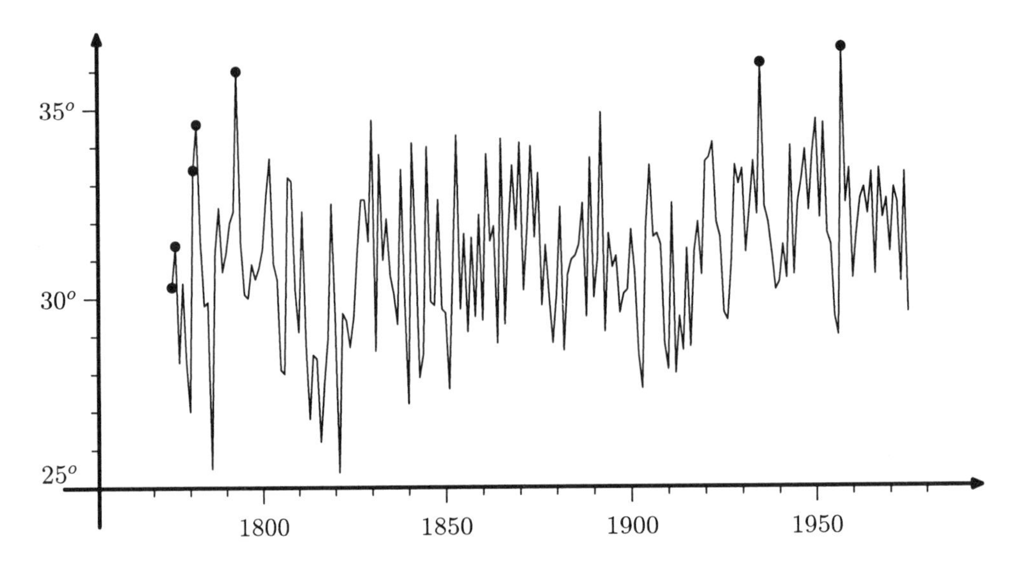

Fig. 4.1 Yearly temperature maxima in Prague.

Proof. We use independence of indicators Y_i. We get

$$\begin{aligned}
&\mathsf{P}(W_r = k|N_r) \\
&\quad = \mathsf{P}(Y_i = 0 \text{ for } N_r + 1 \leq i \leq N_r + k - 1,\ Y_{N_r+k} = 1|N_r) \\
&\quad = \frac{N_r}{N_r+1}\frac{N_r+1}{N_r+2}\cdots\frac{N_r+k-2}{N_r+k-1}\frac{1}{N_r+k} \\
&\quad = \frac{N_r}{(N_r+k-1)(N_r+k)}. \qquad \square
\end{aligned}$$

Theorem 4.6 *We have*

$$\mathsf{P}(W_r \leq m|N_r) = \frac{m}{N_r + m}. \tag{4.15}$$

Proof. We have

$$\begin{aligned}
\mathsf{P}(W_r \leq m|N_r) &= \sum_{k=1}^{m} \mathsf{P}(W_r = k|N_r) = \sum_{k=1}^{m} \frac{N_r}{(N_r+k-1)(N_r+k)} \\
&= \sum_{k=1}^{m} \left(\frac{k}{N_r+k} - \frac{k-1}{N_r+k-1} \right) = \frac{m}{N_r+m}. \qquad \square
\end{aligned}$$

Probability (4.15) tends to 1 as $m \to \infty$. We find m such that this probability is for the first time equal to or larger than a given $\alpha \in (0, 1)$. From the equation $m/(N_r + m) = \alpha$, we get

$$m = \frac{\alpha}{1-\alpha} N_r. \tag{4.16}$$

We take the smallest integer that is equal to or larger than the expression on the right-hand side of formula (4.16). We introduce an example. Consider a series $X_1, \ldots, X_{200}$ such that its last term is the rth record. This means that $N_r = 200$. If $\alpha = 0.5$, then we obtain $m = 200$. Thus with probability 0.5 the next record occurs among the next 200 observations, and with probability 0.5, we shall wait for it longer. If we choose $\alpha = 0.95$, which is usual in scientific research for sufficiently large reliability, then $m = 3800$. Such a long time is needed to reasonably ensure that the next record occurs.

Theorems 4.5 and 4.6 concerned conditional distribution of the variable W_r. A formula was also derived for unconditional distribution, namely

$$\mathsf{P}(W_r > m) = \int_0^\infty \frac{1}{(r-1)!}(1 - e^{-x})^m x^{r-1} e^{-x}\, \mathrm{d}x$$

(see Glick 1978).

Now, we mention another interesting result.

Theorem 4.7 *If $r \le m < n$, then*

$$\mathsf{P}(N_{r+1} > n | N_r = m) = \frac{m}{n}.$$

Proof. We have

$$\begin{aligned} \mathsf{P}(N_{r+1} > n|N_r = m) &= \mathsf{P}(W_r > n - m|N_r = m) \\ &= 1 - \mathsf{P}(W_r \le n - m|N_r = m) = 1 - \frac{n-m}{m + (n-m)} = \frac{m}{n}. \quad \square \end{aligned}$$

This implies that

$$\mathsf{P}\left(\frac{N_r}{N_{r+1}} < \frac{m}{n} \middle| N_r = m\right) = \frac{m}{n}.$$

Each number $x \in (0, 1)$ can be approximated arbitrarily closely by a rational number m/n. It was derived (see Glick 1978, Tata 1969), that if $r \to \infty$ and $m/n \to x$ then

$$\mathsf{P}\left(\frac{N_r}{N_{r+1}} < x\right) \to x.$$

It leads to the conclusion that the variable N_r/N_{r+1} has asymptotically rectangular distribution $\mathsf{R}(0, 1)$. Further it was proved (see Glick 1978) that if $r \to \infty$ then for arbitrary positive integer s and arbitrary $x \in (0, 1)$ the relation

$$\mathsf{P}\left(\frac{N_r}{N_{r+s}} < x\right) \to x \sum_{k=0}^{s-1} \frac{(-\ln x)^k}{k!}.$$

holds.

4.9 SOME APPLICATIONS OF RECORDS

On certain highways and roads, vehicles are not allowed to overtake one another (i.e., car A cannot pass car B by moving from one lane of traffic to another). Such parts of roads are sometimes rather long. Then a slow car is usually followed closely ("tailgated") by a queue of vehicles whose drivers wish to go faster. Speed X_i of the ith vehicle can be considered as a random variable, and it is possible to assume that $X_1, X_2, \ldots$ are i.i.d. variables with a continuous distribution. The car going in front of a caravan is slower than all vehicles before it. Its speed is a lower record. The length of the caravan corresponds to interrecord waiting time. Then the number of cars in different caravans increases, and the separation between them also increases. It has been reported in the literature that empirical characteristics correspond quite well to results derived from the theory of records.

Now, a few words about destructive testing are in order. If we want to know under which force a board breaks, we must apply increasing stress until it really breaks. If we have 100 boards and we want to find the weakest of them, it is not necessary to stress most of them to their failure point. We test the first item until it fails. We stop the next test (short of failure) if the second item survives this amount of stress. We continue similarly with the remaining boards. If the boards are checked in random order then the expected number of broken items is only $\mathsf{E}R_{100} = S_{100} = 5.19$. This means that we spare nearly 95% of the boards.

Another application is a solution of a problem regarding the hiring of a secretary (see Section 6.3).

5

Problems that concern waiting

5.1 GEOMETRIC DISTRIBUTION

We start with some theory. Consider independent trials. Let $p \in (0,1)$ be the probability that event A occurs in a trial. This probability is the same in all trials. Then $q = 1 - p$ is the probability that the event A does not occur in a trial. As usual, the occurrence of event A is called *success*. We want to know how many trials must be realized before a success occurs. The probability that the first k trials are failures and the next $(k+1)$th trial is a success, is

$$p_k = pq^k, \qquad k = 0, 1, 2, \ldots. \tag{5.1}$$

It can be easily verified that $\sum p_k = 1$. Distribution (5.1) is called *geometric*. We denote it by $\mathsf{Ge}(p)$. Its expectation is

$$\mu = \sum_{k=0}^{\infty} kp_k = p \sum_{k=1}^{\infty} kq^k.$$

Let $|q| < 1$. Derivative of the formula

$$\sum_{k=0}^{\infty} q^k = \frac{1}{1-q}$$

with respect to q after some simplification gives

$$\sum_{k=1}^{\infty} kq^k = \frac{q}{(1-q)^2}. \tag{5.2}$$

Since $1 - q = p$, we have

$$\mu = \frac{q}{p}. \tag{5.3}$$

If X is a random variable with distribution $\mathsf{Ge}(p)$, then

$$\mathsf{E}X^2 = \sum_{k=1}^{\infty} k^2 p_k = p \sum_{k=1}^{\infty} k^2 q^k.$$

Derivative of formula (5.2) gives

$$\sum_{k=1}^{\infty} k^2 q^k = \frac{q + q^2}{(1-q)^3}. \tag{5.4}$$

From here we obtain

$$\operatorname{var} X = \mathsf{E}X^2 - \mu^2 = \frac{q+q^2}{p^2} - \frac{q^2}{p^2} = \frac{q}{p^2}. \tag{5.5}$$

We introduce an important formula for the geometric distribution. Let m be a nonnegative integer. Formula (5.1) gives

$$\mathsf{P}(X \geq m) = \sum_{k=m}^{\infty} p_k = q^m. \tag{5.6}$$

We illustrate these results on some numerical examples. Consider a geometric distribution with parameter $p = 0.01$. The expectation of this distribution according to (5.3) is $q/p = 0.99/0.01 = 99$. But if we insert $m = 69$ into (5.6), we obtain $\sum_{k=69}^{\infty} p_k = 0.49984$. This sum is nearly 0.5. Since the sum of all probabilities is 1, this means that the first success occurs before the 69th trial with the same probability as after it. This does not contradict the result that the expected number of failures before the first success is 99. The first success can occur with a positive probability after many trials, which explains the larger value of μ.

Let $X \sim \mathsf{Ge}(p)$, where $p \in (0,1)$. Let t and s be positive integers. From the definition of conditional probability and from (5.6), we get

$$\mathsf{P}(X \geq t+s | X \geq t) = \frac{\mathsf{P}(X \geq t+s)}{\mathsf{P}(X \geq t)} = \frac{q^{t+s}}{q^t} = q^s = \mathsf{P}(X \geq s).$$

If we know that no success occurred among the first t trials, then we shall wait for a success with the same chance as if we only started with our trials. As in exponential distribution we say that the geometric distribution is *memoryless*. We show that this is a characteristic property of the geometric distribution, among discrete distributions (see, e.g., Rhodius 1991). Let Y be a discrete random variable which takes values $0, 1, 2, \ldots$ with positive probabilities and satisfies

$$\mathsf{P}(Y \geq t+s | Y \geq t) = \mathsf{P}(Y \geq s), \qquad s, t = 0, 1, 2, \ldots.$$

This yields

$$\mathsf{P}(Y \geq t + s) = \mathsf{P}(Y \geq t)\mathsf{P}(Y \geq s), \qquad s, t = 0, 1, 2, \ldots.$$

Define $f(x) = \mathsf{P}(Y \geq x)$, $x = 0, 1, \ldots$. We know that $f(t+s) = f(t)f(s)$, $s, t = 0, 1, \ldots$. It is clear that $f(0) = 1$. Define $f(1) = q$. Cases $q = 0$ and $q = 1$ are not interesting. Assume $q \in (0, 1)$. We can see that $f(2) = f(1+1) = f(1)f(1) = q^2$. Complete induction gives $f(k) = q^k$, $k = 0, 1, \ldots$. Then

$$\mathsf{P}(Y = k) = \mathsf{P}(Y \geq k) - \mathsf{P}(Y \geq k+1) = q^k - q^{k+1} = (1-q)q^k$$

for $k = 0, 1, \ldots$. We proved that Y has a geometric distribution with parameter $1 - q$.

Using geometric distribution, we can solve several probabilistic problems. We discuss some of them.

Starting with player A, players A and B alternate in making independent trials. A trial is successful with probability p. The player who reaches success first, is the winner. Find probabilities $\mathsf{P}(A)$ and $\mathsf{P}(B)$ that A and B win, respectively. The probability that the first success occurs in the kth trial is $P_k = q^{k-1}p$, $k = 0, 1, \ldots$. This gives

$$\mathsf{P}(A) = \sum_{k=0}^{\infty} P_{2k+1} = \frac{1}{1+q}, \qquad \mathsf{P}(B) = \sum_{k=0}^{\infty} P_{2k+2} = \frac{q}{1+q}.$$

Since $\mathsf{P}(A) + \mathsf{P}(B) = 1$, the play stops after a finite number of trials with probability 1. For $p = 0.5$, we get $\mathsf{P}(A) = \frac{2}{3}$, and for $p = \frac{1}{6}$, we have $\mathsf{P}(A) = \frac{6}{11}$. Formulas for $\mathsf{P}(A)$ and $\mathsf{P}(B)$ clearly show that player A has an advantage. We can ask how large must the probability of success for player B be for probabilities $\mathsf{P}(A)$ and $\mathsf{P}(B)$ to be equal. Let A and B have success in a trial with probabilities p_1 and p_2, respectively. Player A starts. Define $q_1 = 1 - p_1$, $q_2 = 1 - p_2$. In this case we have

$$P_{2k+1} = (q_1 q_2)^k p_1, \qquad P_{2k+2} = (q_1 q_2)^k q_1 p_2,$$

so that

$$\mathsf{P}(A) = \sum_{k=0}^{\infty} P_{2k+1} = \frac{p_1}{1 - q_1 q_2}, \qquad \mathsf{P}(B) = \sum_{k=0}^{\infty} P_{2k+2} = \frac{p_2 q_1}{1 - q_1 q_2}.$$

Here we also obtain $\mathsf{P}(A) + \mathsf{P}(B) = 1$. Relation $\mathsf{P}(A) = \mathsf{P}(B)$ holds if $p_1 = p_2 q_1$. This gives

$$p_2 = \frac{p_1}{1 - p_1}.$$

If $p_1 = \frac{1}{2}$, then $p_2 = 1$. If $p_1 > \frac{1}{2}$, then the problem has no solution. For $p_1 = \frac{1}{6}$, we get $p_2 = \frac{1}{5}$.

Rabinowitz (1992, p. 286), writes that the next problem was published in the journal *Function* (**6**, 1982/5, p. 33), with its solution in the same journal (**7**, 1983/1, p. 30).

In a contest, n men each toss a cent. If $n-1$ cents agree and the nth does not, the nth takes all the money. If this does not happen, the money jackpots. What on average, does the eventual winner gain?

Let p be probability of a head and $q = 1 - p$ probability of a tail. If $n-1$ cents agree and, the nth does not, we call this event success. The probability of success is

$$P = \binom{n}{1} p^{n-1} q + \binom{n}{1} p q^{n-1} = npq(p^{n-2} + q^{n-2}).$$

Let X be the number of failures before the first success. We know that X has the geometric distribution with parameter P and its expectation is $\mathsf{E}X = (1-P)/P$. If there are X failures before the first success, then the winner gains $nX + n$ cents. The expected gain is

$$\mathsf{E}(nX + n) = n\frac{1-P}{P} + n = \frac{n}{P}.$$

If the cents are fair, then $p = q = \frac{1}{2}$ and the expected gain is 2^{n-1}. However, the winner must wait for such a large amount of money for a long time. The expected number of rounds needed for success is

$$\mathsf{E}(X+1) = \frac{1}{P} = \frac{2^{n-1}}{n}.$$

Consider again the original formulation of the problem. Sometimes we count the number of failures before the rth success. The case $r = 1$ was analyzed above. Here we assume that r is a positive integer. If the rth success occurs in the nth trial, we have $n \geq r$. Then it is natural to write n in the form $n = r + k$, where k is the number of failures before the rth success. It implies that there are $(r-1)$ successes and k failures among the first $n-1$ trials (in arbitrary order) and immediately after that success occurred in the nth trial. The probability of such a result is

$$p_{k,r} = \binom{r+k-1}{k} p^r q^k, \qquad k = 0, 1, 2, \ldots. \tag{5.7}$$

Formula (5.7) can be rewritten as

$$p_{k,r} = \binom{-r}{k} p^r (-q)^k, \qquad k = 0, 1, 2, \ldots. \tag{5.8}$$

In view of (5.8), the distribution is called a *negative binomial*. The Taylor series for $(1-q)^{-r}$ is

$$(1-q)^{-r} = \sum_{k=0}^{\infty} \binom{-r}{k} (-q)^k,$$

and so

$$\sum_{k=0}^{\infty} p_{k,r} = 1 \tag{5.9}$$

for arbitrary $r = 1, 2 \ldots$. But formulas (5.7) to (5.9) are valid for each positive number r. We use the following terminology. If r is an arbitrary positive number, we have *negative binomial distribution* $\mathsf{NBi}(r, p)$. If r is a positive integer, we speak about *Pascal distribution*. We already know that the geometric distribution arises when $r = 1$.

We derive expectation μ and variance σ^2 of the negative binomial distribution. Since $p_{k,r}$ are probabilities, we have

$$\sum_{k=0}^{\infty} \binom{r+k-1}{k} p^r (1-p)^k = 1.$$

The derivative with respect to p gives

$$\sum_{k=0}^{\infty} k \binom{r+k-1}{k} p^r (1-p)^k = r \frac{1-p}{p}. \tag{5.10}$$

The second derivative of (5.10) is

$$\frac{r}{p} \sum_{k=0}^{\infty} k \binom{r+k-1}{k} p^r (1-p)^k - \frac{1}{1-p} \sum_{k=0}^{\infty} k^2 \binom{r+k-1}{k} p^r (1-p)^k = -\frac{r}{p^2}.$$

Using (5.10), we get

$$\sum_{k=0}^{\infty} k^2 \binom{r+k-1}{k} p^r (1-p)^k = \frac{(1-p) r (rq+1)}{p^2}. \tag{5.11}$$

If X is a random variable with distribution $\mathsf{NBi}(r, p)$, then (5.11) yields

$$\mu = \mathsf{E}X = \frac{rq}{p}.$$

Expression (5.11) gives $\mathsf{E}X^2$. Since $\sigma^2 = \operatorname{var} X = \mathsf{E}X^2 - (\mathsf{E}X)^2$, we obtain

$$\sigma^2 = \operatorname{var} X = \frac{rq}{p^2}.$$

5.2 PROBLEM ABOUT KEYS

A woman has a bunch of n keys. Exactly one of them fits a lock. She randomly chooses a key to see if it fits. If not, she returns the key to the bunch, then tries another key, then another one, and so on. Find the expected number of trials until the lock is open.

We have a series of independent trials. The choice of the correct key is called success. The probability of the success is $p = 1/n$. The number of failures before the first success has the geometric distribution. Its expectation according to formula (5.3) is

$$\frac{1-p}{p} = \frac{1-\frac{1}{n}}{\frac{1}{n}} = n - 1.$$

Since the number of trials also includes the successful one, the expected number of trials needed to open the lock is n.

We solve similar problem, except that the tested keys are not returned back to the bunch. In this case all $n!$ orderings of keys are equally probable. The number of ways such that the correct key is on the ith place ($i = 1, \ldots, n$) is $(n-1)!$. Probability, that the correct key is chosen in the i-th trial, is

$$p_i = \frac{(n-1)!}{n!} = \frac{1}{n}.$$

The expected number of trials needed for finding the correct key is

$$\sum_{i=1}^{n} ip_i = \frac{1}{n}(1 + \cdots + n) = \frac{n+1}{2}.$$

5.3 COLLECTION PROBLEMS

A manufacturer producing sweets puts a photograph of a movie star into each box of sweets. There are 20 different photos. Assume that the number of each photo is the same and that the boxes are sold in random order. A customer who collects all 20 different photos receives a prize. Find the expected number of boxes that the customer must buy to have a complete collection of photos.

Assume that the number of boxes is very large and there are n different photos. Let S be the number of boxes needed for a complete collection. Define X_i as the number of boxes bought after having i different photos, until a box with a new photo is bought. We can write

$$S = 1 + X_1 + X_2 + \cdots + X_{n-1}.$$

Of course, X_i is the number of trials until the success occurs. The probability of success (i.e., that a box with a new different photo is bought) is $(n-i)/n$. Formula (5.3) gives only the number of failures: we must increase this number by 1 for the purchase of the box with a new photo. This gives

$$\mathsf{E}X_i = 1 + \frac{i}{n-i} = \frac{n}{n-i},$$

so that

$$\mathsf{E}S = n\left(\frac{1}{n} + \frac{1}{n-1} + \cdots + \frac{1}{2} + 1\right).$$

If $n = 20$, then $\mathsf{E}S = 72.0$, but for $n = 40$, we obtain $\mathsf{E}S = 171.1$ and for $n = 100$ (if we are still interested in it), we have $\mathsf{E}S = 518.7$. Using the Euler formula (4.4)

Table 5.1 Variance of S.

n	20	100
Exact var S	566.5	15,831
Rough approximation var S	658.0	16,449
Fine approximation var S	558.0	15,932

instead of exact calculation, we obtain for values $n = 20, 40, 100$ results $\mathsf{E}S \doteq 71.5$, 170.6, 518.2, respectively.

We also calculate the variance of variable X. Since X_i are independent, we obtain

$$\operatorname{var} S = \sum_{i=1}^{n-1} \operatorname{var} X_i = n \sum_{i=1}^{n-1} \frac{i}{(n-i)^2}. \tag{5.12}$$

Further we investigate limit behavior of var S as $n \to \infty$. It can be easily verified that

$$\operatorname{var} S = n \sum_{i=1}^{n-1} \left[\frac{n}{(n-i)^2} - \frac{1}{n-i} \right].$$

It gives

$$\begin{aligned} \lim_{n\to\infty} \frac{1}{n^2} \operatorname{var} S &= \lim_{n\to\infty} \left[1 + \frac{1}{2^2} + \cdots + \frac{1}{(n-1)^2} \right] \\ &\quad - \lim_{n\to\infty} \frac{1}{n} \left[1 + \frac{1}{2} + \cdots + \frac{1}{n-1} \right]. \end{aligned}$$

The first limit on the right-hand side is $\pi^2/6 \doteq 1.645$, the other limit is zero. We could use a rough approximation

$$\operatorname{var} S \doteq \frac{n^2\pi^2}{6}. \tag{5.13}$$

To make our method more exact, we could use the Euler formula (4.4). In this case we obtain a new approximation (let us say a fine approximation)

$$\operatorname{var} S \doteq \frac{n^2\pi^2}{6} - n[\ln(n-1) + \gamma],$$

where $\gamma = 0.577\ldots$. Numerical results are listed in Table 5.1. We have chosen values $n = 20$ and $n = 100$.

If a man buys r boxes of sweets at once ($r \geq n$), the probability that he has a complete collection of photos is

$$P_{r,n} = \sum_{i=0}^{n} (-1)^i \binom{n}{i} \left(1 - \frac{i}{n}\right)^r.$$

This result can be derived using the *method of inclusion and exclusion*. We discuss this method in Section 7.9. The reader can find formula $P_{r,n}$ with a detailed derivation in Feller [1968, Vol. I, Chap. IV, number (2.3)].

5.4 WHEN TWO PLAYERS WAIT FOR A SUCCESS

Two players, say, A and B, simultaneously conduct independent trials with probability of success p. Each player in her series waits for success. As soon as she reaches it, she must wait for her colleague, who was not so lucky. At the moment that the other player also reaches success, the play stops. Let Z be the length of the play. Find the expectation $\mathsf{E}Z$.

Let X and Y be the number of trials that A and B need for the first success, respectively. It is clear that $Z = \max(X, Y)$ and that X and Y are independent. Define $\xi = X - 1, \eta = Y - 1, \zeta = \max(\xi, \eta)$. Then ξ and η have distribution $\mathsf{Ge}(p)$. We have $Z = \zeta + 1$. Since $\zeta \le j$ holds if and only if $\xi \le j$ and in the same time $\eta \le j$, we obtain for $j = 0, 1, 2, \ldots$ that

$$\mathsf{P}(\zeta \le j) = \mathsf{P}(\xi \le j)\mathsf{P}(\eta \le j) = \left(\sum_{k=0}^{j} pq^k\right)\left(\sum_{l=0}^{j} pq^l\right) = (1 - q^{j+1})^2.$$

If $j = 1, 2, \ldots$ then we obtain

$$\begin{aligned} \mathsf{P}(\zeta = j) &= \mathsf{P}(\zeta \le j) - \mathsf{P}(\zeta \le j - 1) = (1 - q^{j+1})^2 - (1 - q^j)^2 \\ &= pq^j(2 - q^j - q^{j+1}). \end{aligned}$$

Further, we have $\mathsf{P}(\zeta = 0) = \mathsf{P}(\zeta \le 0) = p^2$. Formula (5.2) yields

$$\mathsf{E}\zeta = \sum_{j=0}^{\infty} j\mathsf{P}(\zeta = j) = \frac{q(2+q)}{p(1+q)}.$$

Since $Z = \zeta + 1$, the expectation is $\mathsf{E}Z = \mathsf{E}\zeta + 1$, or

$$\mathsf{E}Z = \frac{1 + 2q}{1 - q^2}.$$

For example, if $p = 0.01$, then A waits for her first success 100 trials on average. This expectation we know from Section 5.1. In our case for $q = 0.99$ we have $\mathsf{E}Z = 149.7$, so that players A and B will wait about 50% longer.

It would be worse if players A and B had to wait so long that both have success simultaneously. The probability of such extraordinary success is p^2, and so the expected number of trials needed is $1 + (1 - p^2)/p^2 = 1/p^2$ [see formula (5.3)]. If $p = 0.01$, then this expectation is $10,000$.

5.5 WAITING FOR A SERIES OF IDENTICAL EVENTS

We have a series of independent trials. Each trial has m equally probable outcomes $A_1, \ldots, A_m$. Find the expected number of trials for k consecutive occurrences of at least one of these outcomes.

If $m = 6$ and $k = 10$, we can imagine that we throw a regular dice until a number appears 10 times consecutively. It does not matter which number within $1, \dots, 6$ it is.

Consider a general formulation of the problem. Let E_k be the expectation that we want to find. When the first sequence of $k-1$ consecutive identical outcomes occurs, then the conditional number of additional trials is 1 with probability $1/m$, and E_k with probability $1 - 1/m$. This gives

$$E_k = E_{k-1} + \frac{1}{m} + \left(1 - \frac{1}{m}\right) E_k, \qquad k \geq 2.$$

This formula can be simplified to

$$E_k = mE_{k-1} + 1, \qquad k \geq 2.$$

Since $E_1 = 1$, we obtain

$$E_k = 1 + m + m^2 + \cdots + m^{k-1} = \frac{m^k - 1}{m - 1}$$

(see *Am. Math. Monthly* **86**, 1979, p. 398). In the example with a dice introduced above, namely, for $m = 6$ and $k = 10$, we get $E_{10} = 6,718,464$.

5.6 LUNCH

A group of m women regularly visit a restaurant to have lunch. After lunch one of them is randomly chosen to pay for all. Find the probability that after the kth lunch it happens, for the first time, that one woman must pay repeatedly. Calculate the expected number of lunches until this happens.

Let X be a random variable that is equal to the number of lunches that must take place one woman pays for the second time. It is clear that

$$p_1 = \mathsf{P}(X = 1) = 0, \qquad p_k = \mathsf{P}(X = k) = 0 \text{ for } k \geq m + 2.$$

Consider only the case $2 \leq k \leq m + 1$. The number of all orderings of how to pay for k lunches is m^k. The number of cases when a woman pays repeatedly after the kth lunch is $m(m-1)\dots(m-k+2)(k-1)$. Then

$$p_k = \mathsf{P}(X = k) = \frac{m(m-1)\dots(m-k+2)(k-1)}{m^k}, \quad k = 2, \dots, m+1.$$

Incidentally, the number of sequences of how to pay for k lunches when nobody pays repeatedly is $m(m-1)\dots(m-k+1)$. This implies

$$\mathsf{P}(X > k) = \frac{m(m-1)\dots(m-k+1)}{m^k}, \qquad k = 1, 2, \dots, m+1.$$

From here we can get also probability p_k, since

$$p_k = \mathsf{P}(X = k) = \mathsf{P}(X > k-1) - \mathsf{P}(X > k).$$

Table 5.2 Probabilities $p_k = \mathsf{P}(X = k)$.

k	2	3	4	5	6	7
p_k	0.16667	0.27778	0.27778	0.18519	0.07716	0.01543

If the group has $m = 6$ women, then we obtain the probabilities introduced in Table 5.2.

As for the calculation of expectation, we have

$$\mathsf{E}X = \sum_{k=2}^{m+1} \frac{m!k(k-1)}{m^k(m-k+1)!} = \sum_{j=1}^{m} \frac{m!j(j+1)}{(m-j)!m^{j+1}}.$$

It is not difficult to prove (see *Am. Math. Monthly* **78**, 1971, pp. 1022–1023) that

$$\frac{\mathsf{E}X}{\sqrt{m\pi/2}} \to 1 \qquad \text{as } m \to \infty.$$

First, we write $\mathsf{E}X$ in the form

$$\begin{aligned} \mathsf{E}X &= \sum_{k=1}^{\infty} k\mathsf{P}(X = k) = \sum_{k=0}^{\infty} \mathsf{P}(X > k) \\ &= \sum_{k=0}^{m} \frac{m!}{(m-k)!m^k} = \sum_{j=0}^{m} \frac{m!}{j!m^{m-j}} = \frac{m!e^m}{m^m} \sum_{j=0}^{m} \frac{e^{-m}m^j}{j!}. \end{aligned}$$

But $e^{-m}m^j/j!$ are probabilities of the Poisson distribution with parameter m. The central limit theorem yields

$$\sum_{j=0}^{m} \frac{e^{-m}m^j}{j!} \to \frac{1}{2}.$$

Using the *Stirling formula*

$$n! = \left(\frac{n}{e}\right)^n \sqrt{2\pi n}\left(1 + \frac{1}{12n} + \frac{1}{288n^2} + \cdots\right)$$

we obtain $m!e^m m^{-m}/\sqrt{2\pi m} \to 1$. From here we get the limit property of $\mathsf{E}X$, which is used as an approximation $\mathsf{E}X \doteq \sqrt{m\pi/2}$.

In our case when $m = 6$, we infer directly from Table 5.2 that $\mathsf{E}X = \sum_{k=2}^{7} kp_k = 3.77472$, whereas the approximation $\mathsf{E}X \doteq \sqrt{6\pi/2} = 3.06998$ is rather rough.

Let us introduce another example. A market research institution interviews people who are randomly chosen from $m = 1,000,000$ inhabitants of Prague. Assume that for some reason this sample is with replacement. The expected number of people interviewed before one person will be interviewed a second time is $\mathsf{E}X \doteq \sqrt{1,000,000\pi/2} = 1253$. This is a smaller number than expected.

Table 5.3 Probabilities of arrival of the bus.

i	Interval	$\mathsf{P}(B_i\|A)$	$\mathsf{P}(B_i)$	$\mathsf{P}(B_i\|C_6)$
1	Before 18:00	0.20	0.1960	–
2	18:00 to 18:05	0.25	0.2450	–
3	18:05 to 18:10	0.20	0.1960	–
4	18:10 to 18:15	0.15	0.1470	–
5	18:15 to 18:20	0.10	0.0980	–
6	18:20 to 18:25	0.05	0.0490	–
7	18:25 to 18:30	0.03	0.0294	0.426
8	After 18:30	0.02	0.0196	0.284
$\sum$		1.00	0.9800	0.710

General results can be also applied to a variant of the well-known *birthday problem*. Many people (more than 365) meet in a large hall. One after the other announces his or her birthday. We neglect the case that a person can have birthday on February 29th and assume that the birthday can be with the same probability 1/365 on any day in the year and that the birthdays of people in the hall are independent of one another. Let X be the random variable that describes the number of people who announced their birthdays until the birthday begins to appear for more than one person. In this case

$$\mathsf{E}X \doteq \sqrt{365\pi/2} = 23.94.$$

5.7 WAITING STUDENT

A student lives in a dormitory in Prague, but he visits his parents every weekend. Sunday evenings he returns to Prague by a bus that is due at 18:00 o'clock (6:00 p.m.). The student gets in at a small stop that is rather far from the terminal. Because of traffic and weather conditions, the arrival of the bus is not exact. In some extraordinary cases a bus may have a defect or accident and it may not arrive at all.

The student arrived at the stop sufficiently early to ensure that he will not be late. But time passes, it is 18:25 o'clock (6:25 p.m.) and the bus has not arrived. The student is naturally interested in the probability that the bus arrives during next 5 minutes. He is even more interested in probability that the bus comes at all.

We can divide the time interval around 18 o'clock into k intervals. These intervals are introduced in the second column of Table 5.3. In our case we chose $k = 8$. Let B_i be the event that the bus arrives in the ith interval ($i = 1, \ldots, k$). Let $A = \cup_{i=1}^{k} B_i$ be the event that the bus comes at all. The complementary event A^c is that the bus does not come. Define B_i the event that the bus arrives in the ith interval ($i = 1, \ldots, k$). Since the student has traveled for a few years, he knows from his experience that $\mathsf{P}(A) = 0.98$. Similarly, he defines from his experience the conditional probabilities $\mathsf{P}(B_i|A)$ introduced in the third column of Table 5.3. Here $\mathsf{P}(B_i|A)$ is the probability that the bus arrives in the ith interval given that it comes at all. The next column of the table contains unconditional probabilities $\mathsf{P}(B_i)$. Since $B_i \subset A$, we have $\mathsf{P}(B_i|A)\mathsf{P}(A) = \mathsf{P}(B_i \cap A) = \mathsf{P}(B_i)$.

Let C_j $(1 \leq j < k)$ be the event that the bus does not arrive in the first j intervals. These events are disjoint, and so we get

$$\mathsf{P}(C_j) = \sum_{i=j+1}^{k} \mathsf{P}(B_i) + \mathsf{P}(A^c) = 1 - \sum_{i=1}^{j} \mathsf{P}(B_i).$$

In our example we have $j = 6$, hence $\mathsf{P}(C_6) = 0.069$. We calculate conditional probabilities $\mathsf{P}(B_i|C_j)$ for $i = j+1, \ldots, k$. Since $B_i \subset C_j$, we obtain

$$\mathsf{P}(B_i|C_j) = \frac{\mathsf{P}(B_i \cap C_j)}{\mathsf{P}(C_j)} = \frac{\mathsf{P}(B_i)}{\mathsf{P}(C_j)}.$$

These conditional probabilities can be found in the last column of Table 5.3. From here we read answers to our questions.

The probability that the bus arrives from 18:25 to 18:30, given that it did not come until 18:25, is 0.426. The probability that the bus comes today is $\mathsf{P}(B_7|C_6) + \mathsf{P}(B_8|C_6) = 0.710$. And the probability that the bus will not arrive today is 0.290.

5.8 WAITING FOR A BUS IN A TOWN

Assume that a bus in a town should arrive to a given stop hourly on the hour (at 16:00, 17:00, 18:00, etc. hours). The bus never arrives before it is due. It can be late, but this delay never exceeds one hour. Let X_k be a random variable that is equal to the delay of the kth bus. Let $\{X_k\}$ be i.i.d. variables with a continuous distribution having a distribution function F and a density f. It follows from our assumptions that $0 \leq X_k \leq 1$, so that $F(0) = 0$, $F(1) = 1$. Denote T_x as the waiting time for a passenger who comes to the stop at time x after 12:00 noon. Without loss of generality, assume that $0 < x < 1$. The probability, that the passenger missed the bus that was due at noon is $F(x)$. We calculate the distribution function of the waiting time of this passenger.

Theorem 5.1 *We have*

$$\mathsf{P}(T_x < t) = \begin{cases} F(t+x) - F(x) & \text{for} \quad 0 \leq t < 1 - x, \\ 1 - F(x) + F(x)F(t + x - 1) & \text{for} \quad 1 - x \leq t < 2 - x, \\ 1 & \text{for} \quad 2 - x \leq t. \end{cases}$$

The corresponding density is

$$g_x(t) = \begin{cases} f(t+x) & \text{for} \quad 0 < t < 1 - x, \\ F(x)f(t + x - 1) & \text{for} \quad 1 - x < t < 2 - x, \\ 0 & \text{otherwise.} \end{cases}$$

Proof. If $t + x < 1$, then $\mathsf{P}(T_x < t)$ is the probability that the passenger will board the bus that was due at noon. It is the probability that the delay will be smaller than $t + x$ but larger than x. This gives $\mathsf{P}(T_x < t) = F(t + x) - F(x)$.

Consider t such that $1 < t + x < 2$. Then $\mathsf{P}(T_x > t)$ is the probability that the noon bus has already left and the next bus will arrive later than at time $x + t$. Then $\mathsf{P}(T_x > t) = F(x)[1 - F(t+x-1)]$ and $\mathsf{P}(T_x < t) = 1 - F(x)[1 - F(t+x-1)]$.

Derivation with respect to t gives the density. □

Theorem 5.2 *Let* $\mu = \mathsf{E}X_k$, $\mu_2' = \mathsf{E}X_k^2$ *and* $\sigma^2 = \mathsf{var}\, X_k = \mu_2' - \mu^2$. *Then*

$$\int_0^1 F(u)\,\mathrm{d}u = 1 - \mu$$

and

$$\mathsf{E}T_x = (\mu + 1 - x)F(x) + \int_0^{1-x} tf(t + x)\,\mathrm{d}t.$$

Proof. We derive the first formula using integration by parts, because

$$\int_0^1 F(u)\,\mathrm{d}u = [uF(u)]_0^1 - \int_0^1 uf(u)\,\mathrm{d}u = 1 - \mu.$$

Further, we can see that

$$\mathsf{E}T_x = \int tg_x(t)\,\mathrm{d}t = \int_0^{1-x} tf(t + x)\,\mathrm{d}t + \int_{1-x}^{2-x} tF(x)f(t + x - 1)\,\mathrm{d}t.$$

Since $\int tf(t + x - 1)\,\mathrm{d}t = tF(t + x - 1) - \int F(t + x - 1)\,\mathrm{d}t$, we have

$$\begin{aligned}
\int_{1-x}^{2-x} tf(t + x - 1)\,\mathrm{d}t &= [tF(t + x - 1)]_{1-x}^{2-x} - \int_{1-x}^{2-x} F(t + x - 1)\,\mathrm{d}t \\
&= (2 - x)F(1) - (1 - x)F(0) - \int_0^1 F(u)\,\mathrm{d}u \\
&= 2 - x - (1 - \mu) = \mu + 1 - x. \quad \square
\end{aligned}$$

Consider the special case when the arrival of the bus has rectangular distribution $\mathsf{R}(0, 1)$. Then $F(x) = x$ and $f(x) = 1$ for $x \in (0, 1)$. Inserting terms into Theorem 5.1, we obtain density of the waiting time for the bus:

$$g_x(t) = \begin{cases} 1 & \text{for} \quad 0 < t < 1 - x, \\ x & \text{for} \quad 1 - x < t < 2 - x. \end{cases}$$

(Verify by integration that it is really a density.) Theorem 5.2 implies that

$$\mathsf{E}T_x = \frac{1 + x - x^2}{2}.$$

This gives $\mathsf{E}T_0 = \mathsf{E}T_1 = \frac{1}{2}$. Function $\mathsf{E}T_x$ reaches its maximum at the point $x = \frac{1}{2}$, and this maximum is $\frac{5}{8}$. This means that even when the buses arrive completely randomly, it is worthwhile to arrive at the stop exactly according to the timetable, because then the expected waiting time is shortest and it makes 30 minutes. The least favorable scenario is to come in the middle between two consecutive arrivals according to the timetable. Then the expected waiting time is $\frac{5}{8}$ of an hour, which is 37 minutes and 30 seconds.

Consider again the general case when the delay of the bus has a general distribution function F. Assume, however, that the moment at which the passenger arrives at the stop is random.

Theorem 5.3 *Let the moment x, at which the passenger comes to the stop, be a random variable with the rectangular distribution $\mathsf{R}(0,1)$. Let σ^2 be variance corresponding to the distribution function F. Then the expectation of the waiting time for the next bus is $\frac{1}{2} + \sigma^2$.*

Proof. Integration by parts, as in the case of the first formula in Theorem 5.2, gives

$$\int_0^1 uF(u)\,\mathrm{d}u = \frac{1-\mu_2'}{2}.$$

A generalization of the theorem of total expectation, which can be found in textbooks on probability theory, gives

$$\begin{aligned}
\mathsf{E}T &= \mathsf{E}(\mathsf{E}T_x) = \int_0^1 \mathsf{E}T_x\,\mathrm{d}x \\
&= \int_0^1 (\mu+1-x)F(x)\,\mathrm{d}x + \int_0^1 \left(\int_0^{1-x} tf(t+x)\,\mathrm{d}t\right)\mathrm{d}x \\
&= (\mu+1)\int_0^1 F(x)\,\mathrm{d}x - \int_0^1 xF(x)\,\mathrm{d}x + \int_0^1 \left(t\int_0^{1-t} f(t+x)\,\mathrm{d}x\right)\mathrm{d}t \\
&= (\mu+1)(1-\mu) - \int_0^1 xF(x)\,\mathrm{d}x + \int_0^1 t[F(1)-F(t)]\,\mathrm{d}t \\
&= \frac{1}{2} + \sigma^2. \quad \square
\end{aligned}$$

If the bus arrives exactly according to the timetable, the expected waiting time is $\frac{1}{2}$ for the case that the passenger comes randomly. If the bus also arrives randomly, then this expectation is larger by a term σ^2.

This problem is introduced in the second volume of the well-known book by Feller (1968).

6

Problems that concern optimization

6.1 ANALYSIS OF BLOOD

A large number N of people must be examined if they suffer from a disease D. These examinations can be organized using one of the following methods:

1. The blood of each person is analyzed separately. It is clear that in this case N analyses of blood are needed.

2. A mixture of blood of k people is analyzed. If the result is negative, then nobody from this group suffers from D. If the result is positive, the blood of each person from this group must be, in addition, analyzed separately.

We would like to know if method 2 is better than method 1. If the answer is positive, we want to find optimal size k of the group.

Assume that people are selected in random order for *blood examination*. Let p be the probability that a randomly chosen person suffers from D. Define $q = 1 - p$. Then the probability that nobody from k people in a group suffers from D is q^k. Assume, for simplicity, that N/k is an integer. In such a case this integer represents a number of groups.

Let X_i be the number of analyses that must be executed for people belonging to the ith group ($i = 1, 2, \ldots, N/k$). Then $X = \sum X_i$ is the total number of analyses. We have

$$X_i = \begin{cases} 1 & \text{with probability } q^k, \\ 1 + k & \text{with probability } 1 - q^k. \end{cases}$$

Table 6.1 Argument minimizing function $f(x)$.

q	Minimum	Argument
0.9	0.593	3.8
0.99	0.195	10.5
0.999	0.063	32.1
0.9999	0.020	100.5
0.99999	0.006	316.7
0.999999	0.002	1000.9

This implies that

$$\begin{aligned} \mathsf{E}X_i &= q^k + (1+k)(1-q^k) = 1 + k(1-q^k), \\ \mathsf{E}X &= \frac{N}{k}[1 + k(1-q^k)] = N\left(1 + \frac{1}{k} - q^k\right). \end{aligned}$$

Table 6.1 lists for several values q the minimum of the function $f(x) = 1 + 1/x - q^x$ and the value x, at which the minimum is attained on interval $(0, \infty)$.

For simplicity, we calculated the minimum on the whole interval $[1, \infty)$ and not only on positive integers.

This method obviously works better if probability p is small. For this reason we used values q near 1 in Table 6.1.

If $p = 0.001$, we can see that the optimal size of the group is $k = 32$. In this case it is sufficient (on average) to perform only 6.3% of blood analyses in comparison with the first method. According to the statistical literature, during World War II the United States spared about 80% blood analyses in this way.

It is important to realize that all assumptions of the method are satisfied. In the 1990s there was a widely publicized incident where AIDS test blood samples from many people were so mixed up that the test was insensitive. It is also important to test people in random order, otherwise the method may be not so economical.

This problem is introduced in Feller (1968, Vol. I, Sect. IX.9, Exercise 26). The method based on mixing blood was described in Dorfman (1943).

Consider another variant of this problem. Assume that $N = m^2$ is the square of an integer and it is known that there is exactly one individual among N persons who suffers from disease D. All N persons can be this one individual with the same probability. We form groups with k people. We restrict our study to such k for which $j = N/k$ is an integer. If blood analysis in a group is positive, blood samples from all people of this group are analyzed separately until the ill (D-afflicted) person is identified. If the ill person is the last in the group, his or her blood is also analyzed for a check, although it is not necessary under our assumptions.

Let X be the number of analyses needed for identification of the ill person. Let Y be the random variable that indicates in which group the ill person is found. Then Y takes values $1, 2, \ldots, j$ with the same probability. The theorem of total expectation gives

$$\begin{aligned}\mathsf{E}X &= \sum_{s=1}^{j} \mathsf{E}(X|Y=s)\mathsf{P}(Y=s) = \sum_{s=1}^{j}\left(s+\frac{1+k}{2}\right)\frac{1}{j} \\ &= \frac{1}{j}\frac{j(j+1)}{2} + \frac{1+k}{2} = 1 + \frac{j+k}{2}.\end{aligned}$$

The sum of positive numbers $x+y$ under the condition $xy = N$ is minimal if $x = y = \sqrt{N}$. Since we assumed that $N = m^2$, numbers $x = y = m$ are integers. Hence the expected number of analyses of blood is minimal in the case $j = k = m$.

6.2 OVERBOOKING AIRLINE FLIGHTS

Each airplane of a commercial airline company can take k passengers. Customers can book tickets in advance, but they pay only before departure at the airport. A customer who has booked a ticket comes to the airport only with probability p. The price of the ticket is d. Since not all customers who have booked reservations on the plane, come to the airport, the company decided to book more seats than the plane contains. If these additional passengers are "bumped" from the flight, the company must compensate those who must wait for the next flight. Let n be the number of booked seats. We consider $n \geq k$. Three cases are analyzed in Austin (1982).

1. Booking is free of charge. If a customer who booked a seat comes to the airport and finds the plane full, he or she would travel on the next plane free of charge. Such a customer causes a lost d.

2. Booking is free of charge. A customer who booked a seat comes to the airport but finds the plane to be full; then the company looses d_1, where $d_1 > d$ (the company pays not only for the ticket for the next flight but also for refreshment and accommodation).

3. Let the price for booking one ticket be r. A ticket costs d, and this amount is paid at the airport before departure. A customer who does not find a seat on the plane causes lost d.

The problem is to find the optimal number of bookings such that the gain of the company is maximal.

If customers come to the airport independently and n seats are booked, then j passengers come with probability

$$p_j = \binom{n}{j} p^j (1-p)^{n-j}, \qquad j = 0, 1, \ldots, n.$$

The gain Z of the company is a random variable. Consider case 1 (listed above). If j customers come and $j \leq k$, then the company receives $z_j = jd$ and pays no

Table 6.2 Optimal number of booked seats without charge.

p	0.6	0.8	0.9	0.95
Optimal n	50	37	33	31
$\mathsf{E}Z$	27.25	28.07	28.66	29.04
Q	18	24	27	28.5

compensation. If $j > k$, then the company receives kd but there are $j - k$ customers without seats. They cost $(j - k)d$, and so the gain is only $z_j = kd - (j - k)d = (2k - j)d$. The expected gain of the company is calculated by the formula

$$\mathsf{E}Z = \sum_{j=0}^{n} z_j p_j.$$

This gives

$$\mathsf{E}Z = d\left[\sum_{j=0}^{k} j\binom{n}{j}p^j(1-p)^{n-j} + \sum_{j=k+1}^{n}(2k-j)\binom{n}{j}p^j(1-p)^{n-j}\right].$$

If $n = k$, then the company has no disappointed customers. In this case the expected gain is

$$Q = d\sum_{j=0}^{k} j\binom{k}{j}p^j(1-p)^{k-j} = dkp.$$

To have an idea about $\mathsf{E}Z$, we choose the price of the ticket $d = 1$. This has no influence on the optimal number of booked seats. Assume that the plane's capacity is $k = 30$ customers. If all customers were absolutely reliable and all came (it would correspond to $p = 1$), the company would choose $n = 30$ and its gain would be 30. If $p < 1$, then it cannot have such a gain.

Table 6.2 lists the optimal number of bookings n for several values p and the corresponding maximal expectation $\mathsf{E}Z$. For comparison, the value Q is also introduced.

We can see that even if the customers are not very reliable, probability p is small, and Q is considerably smaller than amount 30, overbooking helps to increase gain nearly to the maximal amount, 30.

In case 2, it can be deduced similarly that the expected gain is

$$\mathsf{E}Z = d\sum_{j=0}^{k} j\binom{n}{k}p^j(1-p)^{n-j} + \sum_{j=k+1}^{n}[k(d+d_1) - jd_1]\binom{n}{j}p^j(1-p)^{n-j}.$$

In case 3, we add gain from the bookings and expected gain from case 1. We then obtain

$$\mathsf{E}Z = nr + d\left[\sum_{j=0}^{k} j\binom{n}{j}p^j(1-p)^{n-j} + \sum_{j=k+1}^{n}(2k-j)\binom{n}{j}p^j(1-p)^{n-j}\right].$$

Table 6.3 Optimal number of payed bookings.

p	0.6	0.8	0.9	0.95
Optimal n	51	38	33	32
$\mathsf{E}Z$	32.28	31.82	31.96	32.17

If we consider again $k = 30$, $d = 1$, and choose $r = 0.1$ (the price of booking is 10% of the price of the ticket), in the case of absolutely reliable customers the company would have a gain of 33 (bookings 3, tickets 30). For some values p we find in Table 6.3 the optimal number of booked seats n and the corresponding expected gain $\mathsf{E}Z$.

6.3 SECRETARY PROBLEM

A manager is looking for a secretary. An advertisement was answered by n people and the manager decided to hire one who was the best typist. The applicants come individually and each of them writes a short dictation. This enables the manager to determine each applicant's typing.

However, each applicant wants to know immediately, whether he or she is hired. Applicants who are refused are no longer interested in this job, leave, and will not return. The problem is to find a mangerial strategy that ensures the highest probability that the best applicant is hired.

Let X_i be the writing speed of ith applicant. Assume that applicants appear in random order and that all the variables X_i are different.

One managerial strategy can be based on the theory of records. Consider $n = 100$, for example. Table 4.3 indicates that in random order of applicants the series $X_1, \ldots, X_n$ has one record with probability $p_{1,100} = 0.01$, two records with probability $p_{2,100} = 0.05$, and so on. The occurrence of five records has the largest probability: $p_{5,100} = 0.21$. If the manager decides to hire the applicant who creates the fifth record, then the probability of hiring the best applicant will be 0.21.

Another manager's strategy is to examine the first $s - 1$ applicants, but to hire none of them. The manager hires the first of the remaining applicants (if that person is present) who has shown better performance than the previous applicants. If such an applicant is not present, the manager must hire the last one. For a given s, we calculate the probability $\pi(s, n)$ that the best applicant is chosen. Finally, we choose s such that $\pi(s, n)$ is maximal.

Theorem 6.1 *For each* $s = 2, 3, \ldots, n - 1$ *we have*

$$\pi(s,n) = \frac{s-1}{n}\left(\frac{1}{s-1} + \frac{1}{s} + \cdots + \frac{1}{n-1}\right).$$

Proof. It follows from our assumptions that all orderings of variables $X_1, \ldots, X_n$ are equally probable. It is clear that

$$\mathsf{P}(X_r = \max\{X_1, \ldots, X_n\}) = \frac{1}{n} \tag{6.1}$$

holds for each $r = 1, \ldots, n$. Let $1 \leq s - 1 < k \leq n$. We prove that

$$\mathsf{P}(\max\{X_1, \ldots, X_{k-1}\} = \max\{X_1, \ldots, X_{s-1}\}) = \frac{s-1}{k-1}. \tag{6.2}$$

Formula (6.2) yields the probability that the maximum of the first $k - 1$ variables $X_1, \ldots, X_{k-1}$ will be on one of the first $s - 1$ places. The probability that the maximum of the first $k - 1$ variables will be on the jth place ($j = 1, \ldots, k - 1$) is $1/(k-1)$ independently of j. It is similar to the derivation of formula (6.1). All these cases are disjoint, and so the theorem of the addition of probabilities leads to formula (6.2).

The conditional probability that the maximum of the first $k - 1$ variables will be on one of the first $s - 1$ places given that $X_k = \max\{X_1, \ldots, X_n\}$ is also $(s-1)/(k-1)$, because all orderings have equal probability. Using (6.2), we see that the probability that both events occur, that is, that the maximum of the first $k - 1$ variables will be on one of the first $s - 1$ places and that X_k will be maximal among all $X_1, \ldots, X_n$, is

$$\frac{s-1}{k-1}\,\frac{1}{n}.$$

But this is the probability that the first value larger than $\max\{X_1, \ldots, X_{s-1}\}$ will be on the kth place and that it will be the value $\max\{X_1, \ldots, X_n\}$. Cases $k = s, s+1, \ldots, n$ are disjoint, and each of them leads to the optimal choice. Thus the probability that the best applicant will be chosen is

$$\pi(s, n) = \frac{s-1}{n} \sum_{k=s}^{n} \frac{1}{k-1}. \qquad \square$$

If $n = 100$, then we can compute that $\pi(s, 100)$ is maximal for $s = 38$ and $\pi(38; 100) = 0.371$. With this strategy the manager examines the first 37 applicants and hires the next one whose typing speed exceeds that of the preceding applicants. The probability that the best applicant will be chosen is 0.371.

If n is large, we can use the Euler formula (4.4). Then

$$\pi(s, n) \doteq \frac{s-1}{n} \ln \frac{n-1}{s-2}. \tag{6.3}$$

Consider the function

$$f(x) = \frac{x-1}{n} \ln \frac{n-1}{x-2}, \qquad 3 \leq x < n.$$

First, we calculate

$$f'(x) = \frac{1}{n}\left(\ln \frac{n-1}{x-2} - \frac{x-1}{x-2}\right).$$

We are looking for x such that $f(x)$ is maximal. From practical reasons we investigate only the case $n \geq 9$. We introduce an auxiliary function:

$$g(x) = (x-2)\exp\left\{\frac{x-1}{x-2}\right\}.$$

We have

$$[\ln g(x)]' = \frac{x-3}{(x-2)^2} > 0 \qquad \text{for} \quad x > 3,$$

so that $\ln g(x)$ is increasing on $[3, \infty)$. Then the function $g(x)$ is also increasing there. We obtain

$$f'(x) > 0 \quad \text{if and only if} \quad 1 + g(x) < n.$$

But $1 + g(3) = 1 + e^2 \doteq 8.389$, and thus $f'(3) > 0$ for all $n \geq 9$. It is also clear that for any given n, we have $1 + g(x) > n$ if x is sufficiently large. Since

$$f''(x) = \frac{1}{n}\frac{3-x}{(x-2)^2},$$

function $f'(x)$ is decreasing on $[3, \infty)$ and it has only one zero point. Function f must reach its maximum at this point. At the same time, x is the zero point of function f' if and only if

$$1 + g(x) = n.$$

We proved already that the function g is increasing. This implies that the zero point of the function f' increases if n increases. We prove that

$$1 + g\left(\frac{n}{3}\right) < n$$

for $n \geq 100$. Actually, we have

$$1 + g\left(\frac{n}{3}\right) = 1 + \frac{n-6}{3}\exp\left\{1 + \frac{3}{n-6}\right\} \leq 1 + \frac{n-6}{3}\exp\left\{\frac{97}{94}\right\} < n - 5 < n.$$

We used the fact that $\exp\{97/94\} \doteq 2.806 < 3$. If $n \geq 100$, then the zero point of the function f' is larger than $n/3$. For comparison, we can calculate that in the case $n = 100$ the root of the equation $f'(x) = 0$ is $x = 37.406$. If n is large, then the root x of equation $f'(x) = 0$ is also large and $(x-1)/(x-2) \doteq 1$. Then $\ln[(n-1)/(x-2)] \doteq 1$ and

$$x \doteq \frac{n-1}{e} + 2. \tag{6.4}$$

The maximum of the function f corresponds to this x. From (6.3) we have

$$\pi(s,n) \doteq \frac{1}{n}\left(\frac{n-1}{e} + 1\right). \tag{6.5}$$

For example, if $n = 100$, then $x \doteq 99/e + 2 \doteq 38.4$ and $\pi(38; 100) \doteq 0.374$. Hence approximations (6.4) and (6.5) are quite good.

The secretary problem has been investigated and generalized in many papers (see Freeman 1983). Our section is based on Mosteller (1965, Problem 47).

6.4 A BIRTHDAY IS NOT A WORKDAY

Once upon a time there was a country with the following law. In each factory and institution in this country people worked every day (including Saturdays and Sundays) except for days when any worker had a birthday. Then the factory celebrated this day and nobody worked. For simplicity we consider only non–leap years having $N = 365$ days. Assume that birthdays are evenly distributed during year and that people are hired in random order. Find the number of workers such that the time devoted to work is maximal (see Mosteller 1965, Problem 34).

Let n be the number of workers in the factory. The probability that they work on the ith day of the year is

$$p_i = \left(1 - \frac{1}{N}\right)^n,$$

since in this case all workers must have their birthdays in the remaining $N - 1$ days. The expected number of people working on the $i$the day of the year is $n(1 - 1/N)^n$. All year round, this makes $\mu(n) = nN(1 - 1/N)^n$. Since

$$\frac{\mu(n+1)}{\mu(n)} = \frac{n+1}{n}\,\frac{N-1}{N},$$

we obtain

$$\mu(1) < \mu(2) < \cdots < \mu(N-1) = \mu(N) > \mu(N+1) > \mu(N+2) > \cdots.$$

A maximum of $\mu(n)$ is reached for $n = N - 1$ as well as for $n = N$ and is equal to

$$\mu(N-1) = \mu(N) = N^2\left(1 - \frac{1}{N}\right)^N.$$

Since also the Nth worker must be paid and that individual's work has no effect, the optimal number of workers is $n = N - 1 = 364$.

It is not known where F. Mosteller found this problem. Since he did not visit Czechoslovakia before 1989, any similarity with our tradition of celebrating birthdays in factories and institutions prior to that time is purely random, or coincidental.

6.5 VOTING

Everyone must make decisions at some time. However, many important circumstances have random character and cannot be exactly forecast. Thus any decision may appear to be either favorable or unfavorable in retrospect after some time. If the decisions are made in similar situations, we can, after some time, estimate the probability p that a given person will make a favorable decision.

Consider the situation when the decision is made by voting. We describe results published in Anděl (1988a). We deal with the simplest case when three persons, denoted as A, B, and C, vote. Assume that A, B, and C make favorable decisions

Table 6.4 List of voting results.

Case	Voting of member A	B	C	Probability	Result of voting
1	G	G	G	$p_1p_2p_3$	G
2	G	G	B	$p_1p_2(1-p_3)$	G
3	G	B	G	$p_1(1-p_2)p_3$	G
4	B	G	G	$(1-p_1)p_2p_3$	G
5	G	B	B	$p_1(1-p_2)(1-p_3)$	B
6	B	G	B	$(1-p_1)p_2(1-p_3)$	B
7	B	B	G	$(1-p_1)(1-p_2)p_3$	B
8	B	B	B	$(1-p_1)(1-p_2)(1-p_3)$	B

with probabilities p_1, p_2, and p_3, respectively. Without loss of generality, assume that

$$p_3 \leq p_2 \leq p_1 \leq 1. \tag{6.6}$$

We calculate the probability that the committee consisting of members A, B, and C makes a favorable decision if each member votes independently. The voting can result in one of eight different scenarios, listed in Table 6.4. The letter G denotes a favorable decision; B, an unfavorable decision.

Since all cases are disjoint, the probability P that the committee makes a favorable decision is

$$\begin{aligned} P &= p_1p_2p_3 + p_1p_2(1-p_3) + p_1(1-p_2)p_3 + (1-p_1)p_2p_3 \\ &= p_1p_2 + p_1p_3 + p_2p_3 - 2p_1p_2p_3. \end{aligned} \tag{6.7}$$

Consider a special case when $p_1 = 0.9$, $p_2 = 0.8$, and $p_3 = 0.75$. Formula (6.7) yields $P = 0.915$, so the committee makes a favorable decision with greater probability than any of its members.

After some time, member C may realize that she makes mistakes most often. To improve the situation, she begins to vote according to member B. The probability of a favorable decision for C increases from $p_3 = 0.75$ to $p_3^* = 0.8$. However, now the voting result is uniquely determined by the decision of B. The probability P^* that the committee makes a favorable decision in this case is 0.8. The loss of independence of C leads to decreasing probability from value 0.915 to value 0.8.

If C tosses a coin instead of thinking, she would have $p_3 = 0.5$. Now formula (6.7) implies that probability P^{**} of a favorable decision by the committee is $P^{**} = 0.85$. We have a rather surprising result that $P^{**} > P^*$.

What would happen if C votes differently from what she considers to be a favorable decision? In such a case we would have $p_3 = 0.25$ and probability of a favorable decision by the committee would be $P^{***} = 0.785$.

When probabilities p_1, p_2, and p_3 are as given above, the loss of independence of member C in the voting leads to nearly the same decrease of total probability P as if she had intentionally decided to do harm.

Table 6.5 Dependence of P on p.

p	0.5	0.6	0.7	0.8	0.9	1
P	0.5	0.648	0.784	0.896	0.972	1

Choose probabilities $p_1 = 0.8, p_2 = 0.7,$ and $p_3 = 0.6$. Then the value $P = 0.788$ is smaller than p_1. In this case it would be better if the committee did not vote and instead followed the decision of member A. It follows from formula (6.7) that in the general case inequality $P > p_1$ holds if and only if

$$p_1 < \frac{p_2 p_3}{(1-p_2)(1-p_3) + p_2 p_3}. \tag{6.8}$$

If A is only a little wiser than his colleagues, then the inequality (6.8) holds and the probability of a favorable decision by entire committee (A, B, and C) is greater than probability p_1, that its wisest member makes a favorable decision.

To what extent can the committee be influenced, by its least wise member, regarding the quality of the decision? Without loss of generality, assume that $0.5 \le p_i \le 1$. If $0 \le p_i < 0.5$ holds for some i, then the ith member can change his or her decision at the last moment and we would have $1 - p_i$ instead of p_i.

Theorem 6.2 *If* $0.5 \le p_3 \le p_2 \le p_1 \le 1$, *then* $P \ge (p_1 + p_2)/2$. *If* $p_3 = 0.5$ *or* $p_1 = p_2 = 1$, *then the equality holds.*

Proof. First, we write P in the form

$$P = p_1 p_2 + p_3(p_1 - p_1 p_2 + p_2 - p_1 p_2).$$

Since

$$p_1 - p_1 p_2 = p_1(1-p_2) \ge 0, \qquad p_2 - p_1 p_2 = p_2(1-p_1) \ge 0, \qquad p_3 \ge 0.5,$$

we get

$$P \ge p_1 p_2 + \frac{1}{2}(p_1 - p_1 p_2 + p_2 - p_1 p_2) = \frac{p_1 + p_2}{2}. \quad \square$$

Finally, consider the case when all three members are equally wise, so that $p_1 = p_2 = p_3 = p$. Then $P = p^2(3 - 2p)$. We find under which conditions inequality $P > p$ holds. This inequality is equivalent to $-p(2p^2 - 3p + 1) > 0$. The left-hand side is $-2p(p-1)(p-\frac{1}{2})$. This expression is positive for each $p \in (\frac{1}{2}, 1)$ and vanishes for $p = 1$ and for $p = \frac{1}{2}$. If $p \in (\frac{1}{2}, 1)$, then the voting leads to a favorable decision with a higher probability than does the opinion of any of its members. When $p = 1$, then all three members, as well as their voting, are infallible. If all three members base their decisions on a toss of a coin (case $p = \frac{1}{2}$), then their voting will have the same meaning as if it were substituted by a single toss. The relationship between p and P can be seen in Table 6.5.

We can easily derive that the difference $P - p$ is maximal for $p = 0.5 + 1/\sqrt{12} \doteq 0.789$. In this case $P = 0.885$.

6.6 DICE WITHOUT TRANSITIVITY

The game of dice has a long tradition. Dice were used in ancient Egypt when the first dynasty reigned (ca. 3100–2613 B.C.). Later on people in ancient Greece and the Roman Empire played with dice. According to a Greek legend it was Palamedeo who invented dice to amuse the bored Greek solders waiting for the battle of Troy.

The earliest book devoted to probability theory was written by Gerolamo Cardano (1501–1576), and it was titled *De Ludo Aleae*. As this title indicates, the book was devoted mostly to dice. The book was published as late as in 1663, about 100 years after it had been written.

In many plays in which dice are used, the player who throws a larger number has an advantage. Perhaps we would prefer such a dice that would have another 6 face instead of a 1 face. But if other players also use such a dice, the advantage would cancel out. However, the possibility of modifying dice leads to interesting problems. Székely (1986) discusses one of them, called the *paradox of transitivity*; see also Anděl (1988/89).

Player A has three dice (say, I, II, and III) with empty faces. She writes numbers on the faces of all three dice. She may use only the usual numbers 1, 2, 3, 4, 5, 6, but any number can be arbitrarily repeated even on the same dice. So A is allowed to create a dice with all faces containing numbers 6 (dots) as well as the classical dice where each number introduced is used only once.

Then player B scrutinizes all three dice and chooses the one that seems best. Player A chooses one of the remaining two dice and then puts aside the remaining dice.

Now, players A and B throw simultaneously, both with their own dice. The player with the larger number is the winner. If both players have the same number, such a result is not taken into account and they throw again.

If player A prepares three identical dice, the play is fair and nobody has any advantage. The question is whether A is able to write numbers such that she has an advantage meaning that she can ensure a larger probability of winning for herself. Intuitively it seems that this is not possible. For example, if dice I were best, player B would choose it since he is allowed to choose first. Theoretically, any difference among the dice would seem to make the situation for player A even worse. But this is not true. Consider the case when player A numbers the dice as indicated in Table 6.6.

First compare dice I and II. As for classical dice, we have $6 \times 6 = 36$ equally probable cases, which are described in Table 6.7. The pairs in which the number on dice I is larger than that on dice II are underlined. There are 15 such pairs. Therefore, the player having dice I beats the player who has dice II with a probability of $\frac{15}{36}$. Hence dice II is better than dice I.

We compare dice II and III similarly (see Table 6.8). Pairs where the number on dice II is larger than that on dice III are underlined. We again have 15 of such pairs. The player having dice II beats the player who has dice III with a probability of $\frac{15}{36}$. Hence dice III is better than dice II.

Table 6.6 Numbering of dice.

Dice I	1 4 4 4 4 4
Dice II	2 2 2 5 5 5
Dice III	3 3 3 3 3 6

Table 6.7 Comparison of dice I and II.

	II					
I	2	2	2	5	5	5
1	1 2	1 2	1 2	1 5	1 5	1 5
4	4 2	4 2	4 2	4 5	4 5	4 5
4	4 2	4 2	4 2	4 5	4 5	4 5
4	4 2	4 2	4 2	4 5	4 5	4 5
4	4 2	4 2	4 2	4 5	4 5	4 5
4	4 2	4 2	4 2	4 5	4 5	4 5

Table 6.8 Comparison of dice II and III.

	III					
II	3	3	3	3	3	6
1	1 2	1 2	1 2	1 5	1 5	1 5
2	2 3	2 3	2 3	2 3	2 3	2 6
2	2 3	2 3	2 3	2 3	2 3	2 6
2	2 3	2 3	2 3	2 3	2 3	2 6
5	5 3	5 3	5 3	5 3	5 3	5 6
5	5 3	5 3	5 3	5 3	5 3	5 6
5	5 3	5 3	5 3	5 3	5 3	5 6

Finally, we compare dice III and I in Table 6.9. We can see that the player having dice III beats the player who has dice I with a probability of $\frac{11}{36}$. Dice I is better than dice III. The quality of dice does not obey the *transitivity law*.

If player A numbers dice in this way, the probability she will win is at least $\frac{21}{36} = 0.583333$. If B chooses any dice, from the remaining two dice, A takes the dice which is better than the dice of player B. It can be proved that $\frac{21}{36}$ is the maximal probability that A can reach. The numbering of dice described above is best possible strategy for player A.

Analyze the following possible objection. Let players A, B, and C have dice I, II and III, respectively. They throw simultaneously. It seems that B wins more frequently than A, C wins more frequently than B, and A wins more frequently than C, which is impossible. The error here is that the result of the play of three players is

Table 6.9 Comparison of dice III and I.

	I					
III	1	4	4	4	4	4
3	3 1	3 4	3 4	3 4	3 4	3 4
3	3 1	3 4	3 4	3 4	3 4	3 4
3	3 1	3 4	3 4	3 4	3 4	3 4
3	3 1	3 4	3 4	3 4	3 4	3 4
3	3 1	3 4	3 4	3 4	3 4	3 4
6	6 1	6 4	6 4	6 4	6 4	6 4

not determined only by comparison of pairs. We must make calculations for the new play with three players. In such a case we have $6 \times 6 \times 6 = 216$ equally probable cases. For example, result (4,5,6) means that 4, 5, and 6 appeared on dice I, II, and III, respectively. With this result, C would be the winner. Detailed tabulation indicates that the probabilities of winning for each players are

$$\mathsf{P}(A) = \frac{75}{216} = 0.346, \qquad \mathsf{P}(B) = \frac{90}{216} = 0.417 \qquad \mathsf{P}(C) = \frac{51}{216} = 0.236.$$

In this variant of the play dice II is the best and dice III is the worst. No controversy arises.

In statistical literature the results are generalized to an arbitrary number n of dice.

6.7 HOW TO INCREASE RELIABILITY

Consider a technical equipment such as an electronic switch. We call it an *element*. If the switch works correctly, then after a "switch on" signal it conducts electric current and after a "switch off" signal it interrupts the current. If the switch is damaged, it remains in either the "switched on" or the "switched off" state and does not respond to signals. We assume that it cannot be damaged in such a way that it would respond to signals in a manner than the way in which i was constructed. If the switch remains permanently in the "switched off" state the damage is interruption. If it is permanently in the "switched on" state, the damage is a short circuit.

In practical applications of this theory, the problem concerns not only switches, but semiconductors and other electronic pieces as well.

We want the element to work for some time. Let p and q be the probability that during this time the element will be damaged by interruption and a short circuit, respectively. Since these events are disjoint, the probability that the element will be damaged is $p + q$. The *reliability* of the element is the probability that it will not be damaged. This probability is $1 - p - q$. It implies that $p + q \leq 1$. We assume that

$$p + q < 1,$$

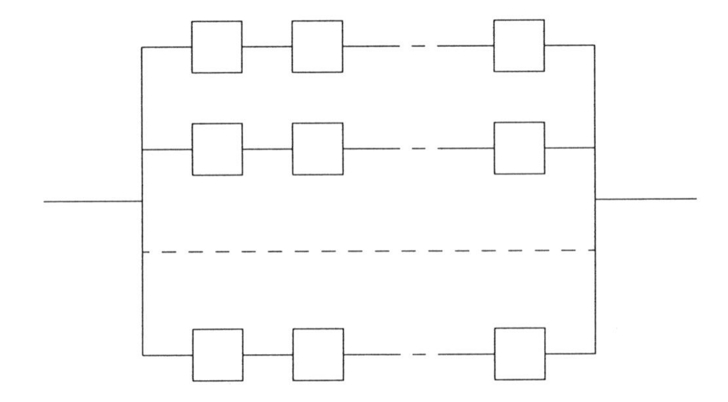

Fig. 6.1 Serial–parallel system.

to ensure a chance that the element will not be damaged. In real situations the elements are quite reliable and p and q are very small positive numbers. In complicated and expensive equipment (e.g., in cosmic rockets) where the reliability must be considerably higher, the problem arises as to how to prepare a more reliable complex from such elements using reserves. One possibility is to use the elements as indicated in Fig. 6.1.

Assume that the elements are ordered in m branches and that each branch has x elements. For our next consideration it is not important that each branch have the same number of elements. But it can be proved that in the optimal system, the maximum difference between the number of elements in any two branches is 1. Details on the proof of this assertion can be found in Kulichová (1994).

The system in Fig. 6.1 will respond correctly to a "switched on" signal if there is at least one branch such that all its elements are switched on simultaneously (some of them can be damaged by a short circuit). The system will respond correctly to a "switched off" signal if all branches are switched off — this means that in each branch at least one element must be switched off.

Assume that all the elements are damaged independently and have the same probabilistic characteristics, that is, the same probabilities p and q. Let

$$w = 1 - p$$

be the probability of the event that an element will not be damaged by interruption. The *reliability of the system* is the probability that the system will work during a given time. The system can work correctly even if some of its elements are damaged. The question of how to construct a sufficiently reliable system from less reliable elements was discussed by technicians and theoreticians in the 1950s (see Moore and Shannon 1956).

First, we calculate the reliability of the system given in Fig. 6.1. Then for a given number of elements x in a branch, we find optimal number of branches, when the reliability of the system is maximal. We further prove that it is theoretically possible to construct a system with reliability arbitrarily near to 1 from elements whose reliability is very low. Finally, we introduce a small numerical study.

Theorem 6.3 *Reliability of the system in Fig. 6.1 is*

$$R(m,x) = (1-q^x)^m - (1-w^x)^m.$$

Proof. The probability that all elements of the first branch are in a short-circuit state, is q^x. The probability that the first branch is not short-circuited is $1-q^x$. The probability that no branch is short-circuited, is $(1-q^x)^m$. The probability that the system is short-circuited, is $\pi_1 = 1-(1-q^x)^m$.

The probability that the first branch is not interrupted is w^x. The probability that this branch is interrupted is $1-w^x$. The system is interrupted when all its branches are interrupted. This event has probability $\pi_2 = (1-w^x)^m$.

Short circuiting of the system and interruption of the system are disjoint events. Hence the probability of damage of the system is $\pi_1+\pi_2$, and its reliability is then $R(m,x) = 1-\pi_1-\pi_2$ □.

Note that the assumption $p+q<1$ implies $q<1-p=w$, so that $(1-q^x)^m > (1-w^x)^m$. Thus we can see that $R(m,x)>0$.

Now, we calculate the optimal value of m. The next theorem was published in Barlow et al. (1963).

Theorem 6.4 *Let x be a given positive integer. Then the maximum of the function $R(m,x)$ taken over integers $m \geq 1$ is reached at $m = \lfloor m_0 \rfloor + 1$, where*

$$m_0 = \frac{x(\ln q - \ln w)}{\ln(1-w^x) - \ln(1-q^x)}$$

and $\lfloor m_0 \rfloor$ is an integer part of number m_0. If m_0 is an integer, then the function $R(m,x)$ reaches maximum at $m = m_0+1$ as well as at $m = m_0$.

Proof. We have

$$\begin{aligned} R(m+1,x) - R(m,x) &= (1-q^x)^{m+1} - (1-w^x)^{m+1} \\ &\quad -(1-q^x)^m + (1-w^x)^m \\ &= -q^x(1-q^x)^m + w^x(1-w^x)^m. \end{aligned}$$

This expression is positive in the case that

$$w^x(1-w^x)^m > q^x(1-q^x)^m.$$

It is easy to see that the last inequality is equivalent to $m < m_0$. The remaining part of the proof is obvious. □

Reliability $R(m,x)$ introduced in Theorem 6.2 contains expressions q^x and w^x. To simplify our discussion, we introduce new parameters M_0 and a such that

$$M_0^{-(1+a)} = q^x, \qquad M_0^{-(1-a)} = w^x.$$

Table 6.10 Number of branches and reliability of the system.

p	q	x	m	$R(m, x)$	M	$R(M, x)$
0.4	0.4	1	2*	0.2	2	0.2
		2	3	0.330 560	4	0.330 099
		3	7	0.447 347	8	0.446 390
		4	15	0.553 058	17	0.549 024
		5	29	0.646 333	35	0.638 682
0.05	0.10	1	1	0.85	3	0.728 875
		2	2	0.970 594	10	0.904 382
		3	4	0.995 592	34	0.966 555
		4	6	0.999 359	110	0.989 060
		5	8	0.999 913	359	0.996 416
0.10	0.05	1	2	0.892 500	4	0.814 406
		2	4	0.988 734	22	0.946 420
		3	7	0.999 018	104	0.987 083
		4	11	0.999 923	493	0.996 923
		5	17	0.999 994	2 327	0.999 273
0.001	0.001	1	2*	0.998	31	0.969 461
		2	3	0.999 997	1 001	0.999 000
		3	4	1.000 000	31 670	0.999 968
		4	6	1.000 000	1 002 003	0.999 999
		5	7	1.000 000	31 701 972	1.000 000

The solution of these two equations is

$$a = \frac{\ln q - \ln w}{\ln q + \ln w}, \qquad M_0 = q^{-x/(1+a)}.$$

We noted above that $q < w$. This implies that $0 < a < 1$. It is obvious that if $x \to \infty$, then $M_0 \to \infty$. Of course, the number M_0 is not the same as m_0 introduced in Theorem 6.4. Also we cannot expect that number $M = \lfloor M_0 \rfloor$ in the same way as optimal number of branches $m = \lfloor m_0 \rfloor$. Nevertheless, the reliability of the system with M branches is so large that it tends to 1 as x grows to infinity.

Theorem 6.5 *If $x \to \infty$, then $R(M, x) \to 1$.*

Proof. We have

$$\begin{aligned} R(M,x) &= \left[1 - M_0^{-(1+a)}\right]^M - \left[1 - M_0^{-(1-a)}\right]^M \\ &= \left[\left(1 - \frac{1}{M_0^{1+a}}\right)^{M_0^{1+a}}\right]^{M/M_0^{1+a}} - \left[\left(1 - \frac{1}{M_0^{1-a}}\right)^{M_0^{1-a}}\right]^{M/M_0^{1-a}}. \end{aligned}$$

If $x \to \infty$ then

$$\frac{M}{M_0^{1+a}} \to 0, \qquad \frac{M}{M_0^{1-a}} \to \infty,$$

$$\left(1 - \frac{1}{M_0^{1+a}}\right)^{M_0^{1+a}} \to e^{-1}, \quad \left(1 - \frac{1}{M_0^{1-a}}\right)^{M_0^{1-a}} \to e^{-1}$$

and we can see that the first term on the right-hand side of the last expression for $R(m, x)$ tends to 1 and the second term tends to zero. □

The optimal number of branches m for some values of p, q, and x is introduced in Table 6.10. In the same table we find the corresponding reliability $R(m, x)$ and values M and $R(M, x)$ for comparison. In the cases denoted by $m = 2^*$ we have $m_0 = 1$ an so the optimal reliability of the system is reached for $m = 2$ as well as for $m = 1$. From practical point of view we would naturally prefer value $m = 1$.

The numbers introduced in Table 6.10 confirm that it is possible to construct quite a reliable system from elements that are not very reliable individually. For example, if $p = q = 0.4$, then the reliability of an element is $R = R(1, 1) = 1 - p - q = 0.2$. But using 29 branches, each of them having five elements, we get a system whose reliability is $R(29, 5) = 0.646$. On the other hand, we cannot continue very far in this way. An excessive number of elements has other disadvantages, such as technical and financial. Table 6.10 also shows that the role of p and q is not symmetric. This was clear from the original formulas. Further, it is evident that parameter M was suitable for our proof but not for practical use. Differences between numbers m and M are in many cases unexpectedly large.

6.8 EXAM TAKING STRATEGY

Some academic tests, especially in the United States, contain a large number of questions. In a limited time a student must answer correctly only a limited number of these questions to achieve a passing score. For example, an applicant to the Society of Actuaries in the United States must pass such a test to attain a fellowship. The applicant must decide how much time to spend on a given question before giving up and moving on to the next question. Chan (1984) proposed a model for such a situation, and O'Brien (1985) generalized it. Assume that the work on an answer consists of two parts. In the first part the problem is screened for S minutes (called *screening time*). After this screening the applicant determines the *working time* W that is needed for preparing a complete answer, such as for computations and writing. Assume that S and W are nonnegative random variables with finite positive expectations $\mathsf{E}S = \gamma$ and $\mathsf{E}W = \nu$ and that (S, W) are independent random vectors for all problems in the test.

A possible strategy for *writing an exam* can be as follows. First, each problem is screened and the working time W for it is determined. If $W > \tau$, no further work will be done and the next problem will be screened. It is necessary to find the optimal value of the *moving-on time* $\tau > 0$. If τ is small, nearly all problems are only screened and the number of solved problems is small. If τ is large, preparing answers to a few difficult questions may take a considerable amount of time and the number of solved problems may also be small. A general solution for determining the optimal τ is given in the next theorem.

Theorem 6.6 *Let W have a density $g(x)$, which is continuous and positive on $(0, \infty)$. Define function*

$$z(\tau) = \frac{\gamma + \int_0^\tau x g(x)\, \mathrm{d}x}{\int_0^\tau g(x)\, \mathrm{d}x}, \qquad \tau > 0. \tag{6.9}$$

Then there exists a unique point $\tau_0 > 0$ such that $\tau_0 = z(\tau_0)$. If in writing exam $\tau = \tau_0$ is chosen, then the expected number of solved problems is maximal.

Proof. Consider a strategy with a time $\tau > 0$. The probability that after screening the problem will be solved is

$$p_\tau = \int_0^\tau g(x)\, \mathrm{d}x. \tag{6.10}$$

The expected working time after screening is

$$\nu_\tau = \mathsf{E}(W | W \le \tau) = \frac{\int_0^\tau x g(x)\, \mathrm{d}x}{\int_0^\tau g(x)\, \mathrm{d}x}, \qquad \tau > 0. \tag{6.11}$$

In n analyzed problems the expected number of worked-out answers is np_τ. The expected time needed for screening and working is $n\gamma + np_\tau \nu_\tau$. This means that one solved problem needs

$$\frac{n\gamma + np_\tau \nu_\tau}{np_\tau} = z(\tau)$$

time, on average.

For $x > 0$, define $G(x) = \int_0^x g(t)\,\mathrm{d}t$. Then G is a positive increasing function on $(0, \infty)$ and $G'(x) = g(x)$ for each $x > 0$. Define $H(\tau) = \int_0^\tau G(x)\,\mathrm{d}x$ for $\tau > 0$. Then H is a continuous increasing function such that

$$\lim_{\tau \to 0+} H(\tau) = 0, \qquad \lim_{\tau \to \infty} H(\tau) = \infty.$$

Integrating by parts, we easily find that for each $\tau > 0$ we have

$$\int_0^\tau xg(x)\,\mathrm{d}x = \int_0^\tau xG'(x)\,\mathrm{d}x = \tau G(\tau) - \int_0^\tau G(x)\,\mathrm{d}x = \tau G(\tau) - H(\tau).$$

This implies that

$$z(\tau) = \frac{\gamma + \tau G(\tau) - H(\tau)}{G(\tau)} = \tau + \frac{\gamma - H(\tau)}{G(\tau)}.$$

We can see that for $\tau \in (0, \infty)$, equality $z(\tau) = \tau$ holds if and only if $H(\tau) = \gamma$. Properties of the function H guarantee that such a point is unique. We denote it by τ_0.

Formula (6.9) gives

$$z'(\tau) = \frac{g(\tau)[\tau - z(\tau)]}{G(\tau)}.$$

If $0 < \tau < \tau_0$, then $H(\tau) < \gamma$. Consequently, $z(\tau) > \tau$, and thus $z'(\tau) < 0$. Similarly, we find that $z'(\tau) > 0$ for $\tau > \tau_0$. This implies that the function $z(\tau)$ attains the global minimum at the point τ_0. □

We mentioned above that two models were introduced in statistical papers. We now describe them. In our analysis we use the well-known formulas

$$\int ye^{-y}\,\mathrm{d}y = -(1+y)e^{-y} + \text{constant},$$
$$\int y^2e^{-y}\,\mathrm{d}y = -(2+2y+y^2)e^{-y} + \text{constant}.$$

Model I. Let W have the density $g(x) = \beta^{-1}e^{-x/\beta}$ for $x > 0$, where $\beta > 0$ is a parameter. Then the variable W has exponential distribution and $\nu = \mathsf{E}W = \beta$. We calculate from (6.9) that

$$z(\tau) = \frac{\gamma + \beta - (\beta + \tau)e^{-\tau/\beta}}{1 - e^{-\tau/\beta}}.$$

The condition $\tau = z(\tau)$ leads to the equation

$$\tau + \beta e^{-\tau/\beta} = \gamma + \beta, \tag{6.12}$$

from which a solution $\tau = \tau_0$ can be calculated numerically. Inserting into (6.10) and (6.11) we get

$$p_{\tau_0} = 1 - e^{-\tau_0/\beta}, \tag{6.13}$$
$$\nu_{\tau_0} = \beta - (\beta + \tau_0)e^{-\tau_0/\beta}. \tag{6.14}$$

Model II. Let W have the density $g(x) = x\delta^{-2}e^{-x/\delta}$ for $x > 0$, where $\delta > 0$ is again a parameter. We obtain $\nu = \mathsf{E}W = \int_0^\infty xg(x)\,\mathrm{d}x = 2\delta$. We further obtain from (6.9)

$$z(\tau) = \frac{\gamma + 2\delta - \left(2\delta + 2\tau + \frac{\tau^2}{\delta}\right)e^{-\tau/\delta}}{1 - \left(1 + \frac{\tau}{\delta}\right)e^{-\tau/\delta}}.$$

In the case $\tau = z(\tau)$, we have the equation

$$\tau + (2\delta + \tau)e^{-\tau/\delta} = \gamma + 2\delta, \tag{6.15}$$

and we can calculate its solution $\tau = \tau_0$ numerically. We insert terms into (6.10) and (6.11), which gives

$$p_{\tau_0} = 1 - \left(1 + \frac{\tau_0}{\delta}\right)e^{-\tau_0/\delta}, \tag{6.16}$$

$$\nu_{\tau_0} = 2\delta - \left(2\delta + 2\tau_0 + \frac{\tau_0^2}{\delta}\right)e^{-\tau_0/\delta}. \tag{6.17}$$

In the abovementioned exam for attaining fellowship in the Society of Actuaries, 60 problems were presented and applicants had a time limit of 3 hours (i.e., 180 minutes). If we assume that the expected time for screening and working out per problem is 6 minutes, then

$$\gamma + \nu = 6.$$

An analysis of model I suggested that $\gamma = 1.77$. Then $\nu = 4.23 = \beta$ and equation (6.12) gives $\tau_0 = 4.561$. Inserting into (6.13) and (6.14), we obtain

$$p_{\tau_0} = 0.660, \qquad \nu_{\tau_0} = 1.239.$$

The expected time needed for 60 problems is

$$60\gamma + 60p_{\tau_0}\nu_{\tau_0} = 155.3,$$

and the expected number of solved problems is

$$60p_{\tau_0} = 39.59.$$

An analysis of model II suggested that $\gamma = 1.4$, so that $\nu = 4.6$ and $\delta = \nu/2 = 2.3$. From (6.15) we obtain $\tau_0 = 4.855$, and inserting into (6.16) and (6.17), we have

$$p_{\tau_0} = 0.623, \qquad \nu_{\tau_0} = 1.625.$$

The expected time needed for 60 problems is

$$60\gamma + 60p_{\tau_0}\nu_{\tau_0} = 144.8$$

and the expected number of solved problems is

$$60p_{\tau_0} = 37.39.$$

The author does not know why different values of γ was used in models I and II.

We add a few concluding remarks. Model I corresponds to the situation when, after screening, the corresponding working time is very short or, at least, not very long. Only in exceptional cases can the working time be longer. On the other hand, a short working time is practically impossible in model II. Here the working time will most often be near to the value τ_1, which maximizes density $g(x)$, specifically, about $\tau_1 = \delta$. Note that this most frequently occurring time is shorter than the expected working time, which is $\nu = 2\delta$. This is because the expected working time is influenced by occasional problems that have rather long working time.

We can also address the issue of how to estimate parameters γ, β, and δ and which models, I or II, to choose. We must remember that these parameters are different for different applicants. In some cases the answer is not difficult. Such exams have been presented for many years and their versions are well-known by now. By analyzing them, we can estimate the parameters and decide between models.

Both the models are rough and approximate. For example, it might be advisable to stop screening if it lasts too long.

6.9 TWO UNKNOWN NUMBERS

A person chooses two different real numbers, say, A and B. This person then each of these numbers writes on a piece of paper and puts both pieces of paper in an urn. Nothing is known about values A and B.

We randomly draw a piece of paper and read the number X written on it. Of course, X is either A or B, each with probability $\frac{1}{2}$. We have two possibilities. We can keep our piece of paper and get a prize equal to the number written on the paper, or we can ask for the other piece of paper and then get the amount written on it.

Numbers A and B can be also negative, in which case we would have to pay such an amount. If we had no choice and if we had to keep the first piece of paper, the expected prize would be $\mathsf{E}X = (A + B)/2$.

The problem is whether we can use information about the number on the paper drawn and increase the probability that our final decision leads to the larger number — and that the expected prize will also be larger.

Ross (1994) recommended the following strategy. Choose a distribution function $G(x)$ increasing on the whole real line. For example, we could take as $G(x)$ the distribution function $\Phi(x)$ of the normal distribution $\mathsf{N}(0, 1)$. We keep the drawn paper with number X with probability $G(X)$ and we ask for the other paper in the urn with probability $1 - G(X)$. Let V be the expected prize in this strategy. Its expectation is

$$\begin{aligned}
\mathsf{E}V &= \mathsf{E}(V|X = A)\mathsf{P}(X = A) + \mathsf{E}(V|X = B)\mathsf{P}(X = B) \\
&= \{AG(A) + B[1 - G(A)]\}\frac{1}{2} + \{BG(B) + A[1 - G(B)]\}\frac{1}{2} \\
&= \frac{1}{2}\{A + B + (B - A)[G(B) - G(A)]\}.
\end{aligned}$$

Since $B - A$ and $G(B) - G(A)$ are nonvanishing numbers with the same sign, their product is positive and $\mathsf{E}V > \frac{1}{2}(A + B)$.

Define $M = \max(A, B)$, $m = \min(A, B)$. Then

$$A + B = M + m, \qquad (B - A)[G(B) - G(A)] = (M - m)[G(M) - G(m)].$$

After elementary arrangement we find that $\mathsf{E}V$ can be written in the form

$$\mathsf{E}V = M\frac{1 + G(M) - G(m)}{2} + m\frac{1 - G(M) + G(m)}{2}.$$

But we also have

$$\mathsf{E}V = M\mathsf{P}(M) + m\mathsf{P}(m),$$

where $\mathsf{P}(M)$ is the probability that we choose a maximum, and

$$\mathsf{P}(m) = 1 - \mathsf{P}(M)$$

is probability that we choose a minimum. Define

$$u = u(M, m) = \frac{1 + G(M) - G(m)}{2}.$$

Then

$$\mathsf{E}V = Mu + m(1 - u) = M\mathsf{P}(M) + m[1 - \mathsf{P}(M)]$$

and

$$M[u - \mathsf{P}(M)] = m[u - \mathsf{P}(M)].$$

If we assumed $u - \mathsf{P}(M) \neq 0$, then we would get $M = m$, which does not hold. Hence $\mathsf{P}(M) = u > \frac{1}{2}$.

Now, consider another variant of the problem (Engel 1996). Let A and B be independent identically distributed random variables with a density f and a distribution function F. Assume that an interval (α, β) is known such that $0 < F(x) < 1$ holds for all $x \in (\alpha, \beta)$. We choose a number $z \in (\alpha, \beta)$. If $X > z$, then we keep the first paper. In the case that $X < z$, we ask for the other paper. If we denote Y as the number on the other paper, then this strategy leads to the choice of the larger number with probability

$$\begin{aligned}
P_1 &= \mathsf{P}(X > z, X > Y) + \mathsf{P}(X < z, X < Y) \\
&= \int_z^{\infty} \int_{-\infty}^{x} f(y)f(x)\,\mathrm{d}y\,\mathrm{d}x + \int_{-\infty}^{z} \int_x^{\infty} f(y)f(x)\,\mathrm{d}y\,\mathrm{d}x \\
&= \int_z^{\infty} F(x)f(x)\,\mathrm{d}x + \int_{-\infty}^{z} [1 - F(x)]f(x)\,\mathrm{d}x \\
&= \left[\frac{1}{2}F^2(x)\right]_z^{\infty} + F(z) - \left[\frac{1}{2}F^2(x)\right]_{-\infty}^{z} \\
&= \frac{1}{2} - \frac{1}{2}F^2(z) + F(z) - \frac{1}{2}F^2(z) \\
&= \frac{1}{2} + F(z)[1 - F(z)].
\end{aligned}$$

It is clear that $P_1 > \frac{1}{2}$. Probability P_1 is maximal in the case when $F(z) = \frac{1}{2}$ (i.e., when z is the *median* of the distribution with the distribution function F). Then we have $P_1 = \frac{3}{4}$.

6.10 ARCHERS

In a contest between two archers, B and C, each has one arrow and they walk together from the origin $x = 0$ to the target at $x = 1$ on the segment $[0, 1]$. Each of them is free to choose the point from which to shoot. Archer B is less accurate than archer C. When both shoot from point x, the probability of hitting the target is x^2 for B and x for C. In the contest, if both hit, then the one that was first to shoot wins. If both hit at the same moment, then both are winners. What is the best plan for B? This problem was given in the Australia Mathematical Olympiad (see Rabinowitz 1992, p. 291).

Archer B can analyze the problem in the following way. Assume that he is with C at point x. If C shot already, there is nothing to solve. Either C hit the target and won, or C did not hit and in this case B can go comfortably to the target and hit it with probability 1. If B and C are at point x and neither of them shot, B has two possibilities. He can shoot at this moment; he then hits with probability x^2. If he does not hit, runs the risk that C may decide to shoot. In this case B wins with probability $1 - x$. If B applies minimax strategy (when he minimizes maximal expected loss), then he shoots if $x^2 \geq 1 - x$. Define $z = (-1 + \sqrt{5})/2 = 0.618$. Then B shoots in the case that $x \geq z$.

Of course, archer C can analyze the same problem similarly. If C shoots at point x, then she wins with probability x. If C does not shoot and B does, then C wins with probability $1 - x^2$. Obviously C decides to shoot in case $x \geq 1 - x^2$. This implies that C also shoots if $x \geq z$.

As soon as B and C reach point $x = z$, if either of them hesitate, the other archer gains an advantage. The optimal strategy for B is to shoot as soon as both B and C come to the point $x = z$.

6.11 A STACKING PROBLEM

We have n books (e.g., n phone books), which we keep in a stack. Let p_i be the probability that the ith book is consulted at the given moment. Assume that $p_i > 0$ for $i = 1, \ldots, n$ and that $p_1 + \cdots + p_n = 1$. After a book is used, it is placed at the top of the stack. Assume that this stacking arrangement was used many times in equidistant moments and that the demands are independent. Let d_i be expected depth of the ith book in the stack.

1. Find the probability P that in a randomly chosen moment (e.g., right now) each book is in its proper place, that is, that the book numbered i is ith from the top, $i = 1, \ldots, n$.

2. Prove that $d_i \leq d_j$ whenever $p_i \geq p_j$. Thus, on the average, the more popular books tends to be closer to the top of the stack.

Problem 1 was formulated in *Mathematics Magazine* (**55**, 1982, p. 300), and its solution appeared in the same journal (**57**, 1984, p. 50). Problem 2 was also published in the same journal (**57**, 1984, p. 175), and its solution appeared later (**58**, 1985, p. 183).

We start with problem 1. The probability that book 1 is on the top of the stack is the probability that it was the last used book, namely, p_1. The probability that book 2 is in place given that book 1 is on the top is

$$p(2|1) = p_2 + p_2 p_1 + p_2 p_1^2 + \cdots = \frac{p_2}{1 - p_1},$$

since book 2 had to appear on the top of the stack in some of the preceding moments, but after that was in demand only book 1. Similarly, we calculate $p(3|1,2)$ and other conditional probabilities, which leads to the formula

$$P = p_1 \frac{p_2}{1 - p_1} \frac{p_3}{1 - p_1 - p_2} \cdots \frac{p_n}{1 - p_1 - p_2 - \cdots - p_{n-1}}.$$

An interesting result can be derived in the case when

$$p_1 = 1 - q, \quad p_2 = q - q^2, \quad \ldots, \quad p_{n-1} = q^{n-2} - q^{n-1}, \quad p_n = q^{n-1},$$

where $q \in (0,1)$. We obtain $P = (1-q)^{n-1}$. This indicates that probability P of the correct order of the book can be arbitrary near 1 if q is sufficiently small.

As for problem 2, we define random variables

$$Y_{kj} = \begin{cases} 1 & \text{if } k\text{-th book is above } j\text{th book,} \\ 0 & \text{otherwise.} \end{cases}$$

Order D_j of the jth book is

$$D_j = 1 + \sum_{\substack{k=1 \\ k \neq j}}^{n} Y_{kj}.$$

Define

$$p_{kj} = \mathsf{P}\{k\text{th book is above } j\text{th book}\}, \qquad k \neq j.$$

Then

$$d_j = \mathsf{E} D_j = 1 + \sum_{\substack{k=1 \\ k \neq j}}^{n} \mathsf{E} Y_{kj} = 1 + \sum_{\substack{k=1 \\ k \neq j}}^{n} p_{kj}.$$

If the kth book is above the jth book, then their relative order is changed if and only if the jth book is consulted. The theorem of total probability gives

$$p_{kj} = p_{kj}(1 - p_j) + p_{jk} p_k = p_{kj}(1 - p_j) + (1 - p_{kj}) p_k = p_k + p_{kj}(1 - p_k - p_j).$$

Table 6.11 Results R of games and their probabilities p.

Game	$R\ p$	$R\ p$	$R\ p$	$R\ p$	$R\ p$	$R\ p$	$R\ p$	$R\ p$
A:B	1 0.1	$\frac{1}{2}$ 0.9	1 0.1	$\frac{1}{2}$ 0.9	1 0.1	$\frac{1}{2}$ 0.9	1 0.1	$\frac{1}{2}$ 0.9
A:C	1 0.2	1 0.2	$\frac{1}{2}$ 0.8	$\frac{1}{2}$ 0.8	1 0.2	1 0.2	$\frac{1}{2}$ 0.8	$\frac{1}{2}$ 0.8
B:C	1 0.6	1 0.6	1 0.6	1 0.6	0 0.4	0 0.4	0 0.4	0 0.4

This implies that

$$p_{kj} = \frac{p_k}{p_k + p_j},$$

so that

$$d_j = 1 + \sum_{\substack{k=1 \\ k \neq j}}^{n} \frac{p_k}{p_k + p_j}.$$

If $p_i \geq p_j$, then $p_k/(p_k + p_i) \leq p_k/(p_k + p_j)$ for $k \notin \{i, j\}$. It is clear that $p_j/(p_j + p_i) \leq p_i/(p_i + p_j)$. Each term in the sum for d_i is less than or equal to the corresponding term in the sum for d_j, and thus $d_i \leq d_j$.

This result can be found also in Lam et al. (1983). So called *move-to-front scheme*, which we used in our problem, is in the long run less efficient than the *transposition scheme* (see Hendricks 1972).

6.12 NO RISK, NO WIN

Three chess players Andy, Bert, and Colin (who will be denoted A, B, and C here), have tied for first place in their chess club tournament. Now, they are to play off for the championship. Each of them is to play one game with both the others. As usual in chess, scoring is 1 for win, $\frac{1}{2}$ for a draw, and 0 for a loss. If their scores are still all level, the round will be repeated. If two of them are level ahead of the third, those two will continue to play until one of them scores a win.

Andy is very cautious. He never looses against either of the others. He never takes risks, and so he has a probability of only 1/10 of beating Bert in any given game, and probability 1/5 of winning against Colin. Games between Bert and Colin are swashbuckling affairs that never result in draws. Bert wins 60% and Colin, 40%. Assume that results of all games are independent, and compare Andy's and Colin's chances of emerging as club champions. Rabinowitz (1992, p. 286), states that the problem was formulated in the journal *Parabola* (**20**, 1984/3, p. 35), and its solution was published in the same journal (**20** 1984/3, p. 36).

The first round consists of three games, which we denote as A:B, A:C, B:C. The result of a game is a number from 1, $\frac{1}{2}$, 0, which describes score of the first named player. For example, result 0 of the game B:C says that in the game played between B and C, player B lost. It follows from the formulation of the problem that 0 can

Table 6.12 Scores of players and their probabilities P.

A	2	1.5	1.5	1	2	1.5	1.5	1
B	1	1.5	1	1.5	0	0.5	0	0.5
C	0	0	0.5	0.5	1	1	1.5	1.5
P	0.012	0.108	0.048	0.432	0.008	0.072	0.032	0.288

never be a result of games A:B and A:C and that $\frac{1}{2}$ can never be a result of game B:C. All possible results R of the three games and their probabilities p are given in Table 6.11.

Scores of players A, B, and C for each of eight possibilities arising from results of three games and their probabilities P are listed in Table 6.12.

Player A will be champion immediately after the first round in cases 1, 3, 5, and 6, the probability of which equals to $0.012 + 0.048 + 0.008 + 0.072 = 0.14$. Player B will be champion in case 4, which has probability 0.432. Player C will be champion in case 8, which has probability 0.288. In case 2 only A and B will continue to play. Since A never looses, sooner or later he will be champion. Similarly, in case 7, only A and C continue and A will be champion. In no case will there be another round for all three players. Then the total probability that A, B, and C will be champions, is $0.14 + 0.108 + 0.032 = 0.28$, 0.432, and 0.288. Adventurer Colin, who loses with Bert more often and never wins with Andy, has a greater probability to be champion than does cautious Andy.

7

Problems on calculating probability

7.1 DORMITORY

The following problem concerns a *college dormitory*. The dormitory was located somewhere in the United States. There were 600 double rooms in the dormitory. At the beginning of the academic year, 1200 students were accommodated in these rooms and all the beds were occupied. It appeared that they were 120 black and 1080 white students and that there were 100 mixed-race roommate pairs (i.e., one student black, the other white). We can easily calculate that 490 rooms were occupied by white pairs of students and 10 by black pairs of students.

Black students complained that the number of *mixed-race rooms* is not large enough and that the assignment of rooms constituted racial discrimination. The management of the dormitory objected that information based on race was not available on application forms or on room assignment records and that pairing was random. Moreover, it seems that the number of mixed rooms is rather large and not small.

Therefore, a statistical analysis should be carried out to determine if 100 mixed-race rooms can be explained by random pairing.

Before solving formulation of the problem presented above, we illustrate the situation by a small example. Assume that there are only three rooms in the dormitory and that four black and two white students ask for accommodation. If b and w denote the fact that a bed is occupied by a black and a white student, respectively, then all possible pairings are introduced in Table 7.1.

We can see that there are 15 possibilities for assigning beds with respect to race. If the pairing is random, then all cases are equally probable. In three cases there were no mixed-race rooms and in the remaining 12 cases two rooms are mixed. Let Z be

Table 7.1 Pairings of black and white students.

$w\,w$	$w\,b$	$w\,b$	$w\,b$	$w\,b$	$b\,w$	$b\,w$	$b\,w$	$b\,w$	$b\,b$	$b\,b$	$b\,b$	$b\,b$	$b\,b$	$b\,b$
$b\,b$	$w\,b$	$b\,w$	$b\,b$	$b\,b$	$w\,b$	$b\,w$	$b\,b$	$b\,b$	$w\,w$	$w\,b$	$w\,b$	$b\,w$	$b\,w$	$b\,b$
$b\,b$	$b\,b$	$b\,b$	$w\,b$	$b\,w$	$b\,b$	$b\,b$	$w\,b$	$b\,w$	$b\,b$	$w\,b$	$b\,w$	$w\,b$	$b\,w$	$w\,w$

a random variable that equals to number of mixed-race rooms. In our small example we have

$$\mathsf{P}(Z=0) = \frac{3}{15} = 0.2, \qquad \mathsf{P}(Z=1) = 0,$$
$$\mathsf{P}(Z=2) = \frac{12}{15} = 0.8, \qquad \mathsf{P}(Z=3) = 0.$$

Then

$$\mathsf{E}Z = 1.6, \qquad \mathsf{E}Z^2 = 3.2, \qquad \operatorname{var} Z = \mathsf{E}Z^2 - (\mathsf{E}Z)^2 = 0.64.$$

Consider a general case. Let R be the number of double rooms in the dormitory. Let W and B be number of white and black students, respectively. The total number of students is $N = W + B$. We assume that all beds in the dormitory are occupied, and so $N = 2R$.

Beds for white students can be determined in $\binom{N}{W}$ ways. If white students have their beds, the pairing with respect to race is completely given. Let z be the number of mixed-race rooms. It is clear that z takes only even values if both W and B are even and takes only odd values if both W and B are odd. Of course, $0 \le z \le \min(W, B)$. We derive the number of ways of pairing students such that they lead to z mixed-race rooms.

We can choose z mixed-race rooms from R rooms in $\binom{R}{z}$ ways. If z rooms are chosen, then in each of them we have two ways to allocate beds for a white student and a black student. Thus we have 2^z ways in z rooms. But $R - z$ rooms remain. We must choose rooms for remaining $W - z$ white students. In other words, we must choose $(W - z)/2$ rooms from remaining $R - z$ rooms. It can be done in

$$\binom{R-z}{\frac{W-z}{2}}$$

ways. All the cases can be combined. Hence, z mixed-race rooms can be chosen in

$$\binom{R}{z} 2^z \binom{R-z}{\frac{W-z}{2}}$$

ways. (Verify this using the example with three rooms introduced above.) If we assume that all cases are equally probable, we get

$$\mathsf{P}(Z = z) = \frac{\binom{R}{z} 2^z \binom{R-z}{\frac{W-z}{2}}}{\binom{N}{W}}.$$

This formula can be written in the form

$$\mathsf{P}(Z = z) = \frac{R!\, 2^z W!\, (N - W)!}{z!\, \frac{W-z}{2}!\, \left(R - z - \frac{W-z}{2}\right)!\, N!}.$$

Let x and y be the number of rooms occupied by pairs of white and black students, respectively. It is clear that

$$x = \frac{W - z}{2}, \qquad y = \frac{B - z}{2} = \frac{N - W - z}{2} = R - z - \frac{W - z}{2}. \tag{7.1}$$

Finally, we obtain

$$\mathsf{P}(Z = z) = \frac{2^z R!\, W!\, B!}{N!\, x!\, y!\, z!}.$$

This formula is introduced in Hader (1967), where the problem was published. Given z, values x and z are uniquely determined by formula (7.1).

At this moment we can determine whether the *pairing of students* can be considered as random. We have $R = 600$, $N = 1200$, $W = 1080$ and $B = 120$. Then Z can be only even. Define

$$S(m) = \sum_{k=0}^{m} \frac{2^{2k} R!\, W!\, B!}{N!\, x!\, y!\, (2k)!}.$$

Let α be a given level of probability (usually, $\alpha = 0.05$). We find numerically the number m_0 such that

$$S(m_0) \leq \alpha < S(m_0 + 1).$$

Define $z(\alpha) = 2m_0$. We call this number the *critical value* on level α. If the real number of mixed-race rooms is smaller than or equal to this critical value, we reject the hypothesis that pairing of students was random. Numerically we obtain $S(49) = 0.0254004$, $S(50) = 0.0598744$, $z(0.05) = 98$. Since the actual number of mixed-race rooms is larger than the critical value, we have no reason to reject the hypothesis that pairing of students was random, without factoring in their race.

If numbers W and B are very large, numerical computation of the probability $\mathsf{P}(Z = z)$ can be very difficult. In some cases an answer can be based on the following procedure. Introduce random variables $Z_1, \ldots, Z_R$ such that

$$Z_i = \begin{cases} 1 & \text{if } i\text{th room is mixed,} \\ 0 & \text{if } i\text{th room is not mixed.} \end{cases}$$

Obviously, the total number of mixed rooms is $Z = Z_1 + \cdots + Z_R$. In the ith room the probability that the first bed is occupied by a white student and the other bed by a black student, is $(W/N)[B/(N-1)]$. Similarly we find that in the ith room the first bed is occupied by a black student and the other bed by a white student with probability $(W/N)[B/(N-1)]$. But $Z_i = 1$ holds if and only if one of these two disjoint events occurs. Therefore,

$$\mathsf{P}(Z_i = 1) = \frac{2WB}{N(N-1)}.$$

Since variable Z_i can be equal only to 0 or 1, we have $\mathsf{E}Z_i = \mathsf{P}(Z_i = 1)$, and so

$$\mathsf{E}Z = R\,\mathsf{E}Z_i = \frac{N}{2}\,\mathsf{E}Z_i = \frac{WB}{N-1}.$$

Let $i \neq j$. Event $\{Z_i = 1\} \cap \{Z_j = 1\}$ occurs in the following four disjoint cases:

	I	II	III	IV
ith room	$w\ b$	$w\ b$	$b\ w$	$b\ w$
jth room	$w\ b$	$b\ w$	$w\ b$	$b\ w$

Probability of the case I is

$$P_{\mathrm{I}} = \frac{W}{N}\frac{B}{N-1}\frac{W-1}{N-2}\frac{B-1}{N-3}.$$

It can be easily verified that $P_{\mathrm{I}} = P_{\mathrm{II}} = P_{\mathrm{III}} = P_{\mathrm{IV}}$. Then $\mathsf{P}(Z_i = 1, Z_j = 1) = 4P_{\mathrm{I}}$. Since the variable $Z_{ij} = Z_iZ_j$ also can be equal only to 0 or 1, we obtain

$$\mathsf{E}Z_iZ_j = \mathsf{P}(Z_iZ_j = 1) = \mathsf{P}(Z_i = 1, Z_j = 1) = 4P_{\mathrm{I}}.$$

This implies

$$\begin{aligned}\mathsf{E}Z^2 &= \mathsf{E}(Z_1 + \cdots + Z_R)^2 = R\,\mathsf{E}Z_i^2 + R(R-1)\,\mathsf{E}Z_iZ_j \\ &= R\frac{2WB}{N(N-1)} + 4R(R-1)\,\frac{W(W-1)B(B-1)}{N(N-1)(N-2)(N-3)}.\end{aligned}$$

Using the relation $2(R-1)/(N-2) = 1$, we have

$$\operatorname{var} Z = \mathsf{E}Z^2 - (\mathsf{E}Z)^2 = \frac{WB}{N-1}\left[1 + \frac{(W-1)(B-1)}{N-3} - \frac{WB}{N-1}\right].$$

(Again, we recommend that you verify that by inserting $R = 3$, $N = 6$, $W = 2$, one gets the result $\mathsf{E}Z = 1.6$, $\operatorname{var} Z = 0.64$.)

Now we use the derived results. For every given $\varepsilon > 0$, *Tshebyshev inequality* gives

$$\mathsf{P}(|Z - \mathsf{E}Z| \geq \varepsilon) \leq \frac{\operatorname{var} Z}{\varepsilon^2}. \tag{7.2}$$

For $N = 1200$, $R = 600$, $W = 1080$, and $B = 120$, we have $\mathsf{E}Z = 108.090$ and $\operatorname{var} Z = 19.348$. If we want to have $\operatorname{var} Z/\varepsilon^2 = \alpha$, we must choose

$$\varepsilon = \sqrt{\frac{\operatorname{var} Z}{\alpha}}.$$

Especially for $\alpha = 0.05$, we obtain $\varepsilon = 19.671$. The probability that Z differs from 108.090 at least by 19.671 is maximally 0.05. Then Z is between 88.4 and 127.8 with probability of at least 0.95. Since in our case Z is even, we proved that the event $\{Z \leq 88\} \cup \{Z \geq 128\}$ has a probability no greater than 0.05.

If the number of mixed-race rooms is 100, we can draw no conclusion from the preceding result. However, if number of such rooms were 85, this would lead to rejection of the hypothesis that pairing was random without any additional calculations.

The procedure based on Tshebyshev inequality rejects the hypothesis that paring was random when the number of mixed-race rooms is either very small or very large. On the other hand, we must admit that procedures based on Tshebyshev inequality are usually very conservative and that the actual probability of the event $\{Z \leq 88\} \cup \{Z \geq 128\}$ is considerably smaller than the chosen level $\alpha = 0.05$.

The problem of how to assign dormitory rooms to students is considerably more complicated. Some students in their applications introduce roommate preferences or request accomodation in specific dormitories, and dormitory management tries to fulfill their wishes. Then the hypothesis of random pairing should be tested using values W and B obtained after eliminating all students for whom such requests were granted. Moreover, our results concern a given academic year. Even if the pairing were completely random, there is a high probability that at least in one academic year the number of mixed-race rooms is small.

7.2 TOO MANY MARRIAGES

A crime occurred in a small country. The jury panel for the criminal trial consists of 156 members selected randomly without replacement from 3000 registered voters. There are 500 *married couples* among those 3000 voters, and the *jury panel* contains M married couples. The defense lawyer argues for a change of venue, asserting that the probability of M or more married couples on the jury panel is less than 1 in 20,000. Find M such that the lawyer's argument is correct. This problem was published in *Am. Math. Monthly* (**96**, 1989, pp. 362–363, Problem E3234).

Let $P(i, j)$ be the probability that the jury panel has exactly

- i married couples
- j people who are married but whose spouse is not selected for jury panel
- $156 - 2i - j$ people who are not married

There are $\binom{500}{i}$ possible ways to choose i married couples from 500 married couples. From remaining $500 - i$ couples, j persons must be chosen in such a way that there are no married couples in this group. We have $\binom{500-i}{j}$ ways to choose j couples, and then one person is chosen from each couple. This makes $\binom{500-i}{j}2^j$ ways altogether. Finally, $156 - 2i - j$ persons must be chosen from 2000 persons, which can be realized in $\binom{2000}{156-2i-j}$ ways. This gives

$$P(i,j) = \frac{\binom{500}{i}\binom{500-i}{j}2^j\binom{2000}{156-2i-j}}{\binom{3000}{156}}.$$

The probability that the jury panel contains M or more married couples is

$$Q(M) = \sum_{i=M}^{78} \sum_{j=0}^{156-2i} P(i,j).$$

We can find numerically that the smallest value M such that $Q(M) \le 1/20,000$, is $M = 8$. More precisely, the probability of eight or more couples is less than 1 in 21,327, and the probability of seven or more couples is more than 1 in 2986. Using methods of advanced probability theory, we can prove (see Tucker 1984, p. 313), that

$$Q(M) = \sum_{j=M}^{78} (-1)^{j-M} \binom{j-1}{M-1} \binom{500}{j} \frac{\binom{156}{2j}}{\binom{3000}{2j}}.$$

However, the problem is not correctly posed from a statistical point of view. One fundamental principle for evaluating data is that the statistical test must be chosen before screening data. It seems that this principle was violated since the defense lawyer attacked the jury panel after having observed too many married couples on it. The defense lawyer could have been satisfied with the number of married couples on the jury panel but have investigated other factors such as age or education structure among panel members with the slogan "One who looks for, finds." An analogous remark is valid for other probability problems mentioned in this book.

7.3 TOSSING COINS UNTIL ALL SHOW HEADS

We have n identical *coins*. The head on each coin occurs with probability p. Assume that $0 < p < 1$. We first toss all the coins. Perhaps heads occur on some of them. We put such coins aside and toss those that show tails after the first toss. Then we toss those that show tails after the second toss, and so on until all coins show heads. It is proved in probability theory that a finite number of tosses suffices with probability 1. Let X_n be the number of coins used in the last toss. Find probability $\mathsf{P}(X_n = i)$, $i = 1, \ldots, n$.

Let $A_i = \{X_n = i\}$ and denote, as usual, $q = 1 - p$. Let B_k be the event that k tosses were needed to complete the experiment. We say that a coin is successful if it showed $(k-1)$-times tail and then in the kth toss head. Probability, that a coin is successful, is $v_k = q^{k-1}p$. A coin is not successful if it showed heads before the kth toss. The probability that a coin is not successful is for $k \ge 2$ equal to

$$w_k = p + qp + q^2p + \cdots + q^{k-2}p = p\frac{1-q^{k-1}}{1-q} = 1 - q^{k-1}.$$

Since $w_1 = 0$, this formula holds for w_k also when $k = 1$. It is clear that

$$\mathsf{P}(A_i \cap B_k) = \binom{n}{i} v_k^i w_k^{n-i} = \binom{n}{i} (pq^{k-1})^i (1 - q^{k-1})^{n-i},$$

and thus

$$\mathsf{P}(X_n = i) = \mathsf{P}(A_i) = \binom{n}{i} p^i \sum_{k=1}^{\infty} q^{i(k-1)} (1 - q^{k-1})^{n-i}. \tag{7.3}$$

Since

$$\begin{aligned}
\sum_{k=1}^{\infty} q^{i(k-1)} (1 - q^{k-1})^{n-i} &= \sum_{k=1}^{\infty} q^{i(k-1)} \sum_{j=0}^{n-i} \binom{n-i}{j} (-1)^j q^{(k-1)j} \\
&= \sum_{j=0}^{n-i} (-1)^j \binom{n-i}{j} \sum_{k=1}^{\infty} q^{(k-1)(i+j)} \\
&= \sum_{j=0}^{n-i} (-1)^j \binom{n-i}{j} \frac{1}{1 - q^{i+j}},
\end{aligned}$$

we have finally

$$\mathsf{P}(X_n = i) = \binom{n}{i} p^i \sum_{j=0}^{n-i} (-1)^j \binom{n-i}{j} \frac{1}{1 - q^{i+j}}.$$

Now, consider expectation $\mathsf{E}X_n$. Formula (7.3) gives

$$\begin{aligned}
\mathsf{E}X_n &= \sum_{i=0}^{n} i\mathsf{P}(X_n = i) = \sum_{k=1}^{\infty} \sum_{i=0}^{n} i \binom{n}{i} (pq^{k-1})^i (1 - q^{k-1})^{n-i} \\
&= \sum_{k=1}^{\infty} (pq^{k-1} + 1 - q^{k-1})^n \\
&\quad \times \sum_{i=0}^{n} i \binom{n}{i} \left(\frac{pq^{k-1}}{pq^{k-1} + 1 - q^{k-1}} \right)^i \left(\frac{1 - q^{k-1}}{pq^{k-1} + 1 - q^{k-1}} \right)^{n-i}.
\end{aligned}$$

We simplify

$$pq^{k-1} + 1 - q^{k-1} = 1 - q^k.$$

Then we realize that

$$\sum_{i=0}^{n} i \binom{n}{i} \left(\frac{pq^{k-1}}{1 - q^k} \right)^i \left(\frac{1 - q^{k-1}}{1 - q^k} \right)^{n-i}$$

is the expectation of a binomial distribution, which corresponds to n trials with probability of the success $pq^{k-1}/(1 - q^k)$. This expectation is $npq^{k-1}/(1 - q^k)$, and so

$$\begin{aligned} \mathsf{E}X_n &= \sum_{k=1}^{\infty}(1-q^k)^n n\frac{pq^{k-1}}{1-q^k} = \frac{n}{q}\sum_{k=1}^{\infty}pq^k(1-q^k)^{n-1} \\ &= \frac{n}{q}\sum_{k=2}^{\infty}pq^{k-1}(1-q^{k-1})^{n-1}. \end{aligned}$$

If $n > 1$, then in view of (7.3)

$$\mathsf{E}X_n = \frac{1}{q}n\sum_{k=1}^{\infty}pq^{k-1}(1-q^{k-1})^{n-1} = \frac{\mathsf{P}(X_n = 1)}{q}.$$

Define $p_n = \mathsf{P}(X_n = 1)$. We have

$$p_n = np\sum_{k=0}^{\infty}q^k(1-q^k)^{n-1}.$$

It can be proved (see *Am. Math. Monthly* **101**, 1994, pp. 78–80, Problem E3436) that the sequence p_n does not converge as $n \to \infty$, but oscillates around $-p/\ln q$. For example, if coins are fair and $p = q = \frac{1}{2}$, then p_n oscillates between 0.721340 and 0.721355 around $1/(2\log 2) = 0.7213475$ without converging.

7.4 ANGLERS

The weights of fish in *Lake Canusa* are rectangularly distributed in the interval (0 kg, 1 kg). This lake is at the border between Canada and the United States, which is expressed in its name, CAN-USA. The both countries have different rules for fishing. Canadians are allowed to fish until they catch a fish that is heavier than the last one. Then they must leave the lake and revisit it another day. Americans are allowed to fish until the total weight is larger than 1 kg. Then they must stop fishing. Compare the rules of both countries. This problem was published in Swift (1983) with a comment that it was presented by R. Engel at The Third International Statistics Institute Round Table on Teaching Statistics.

Weights of fishes can be considered as a sample from the rectangular distribution $\mathsf{R}(0,1)$. Probability C_n that a Canadian catches exactly n fishes is given by the formula

$$\begin{aligned} C_n &= \mathsf{P}(X_1 > X_2 > \cdots > X_{n-2} > X_{n-1} < X_n) \\ &= \int_0^1\int_0^{x_1}\cdots\int_0^{x_{n-3}}\int_0^{x_{n-2}}\int_{x_{n-1}}^1 \mathrm{d}x_n\,\mathrm{d}x_{n-1}\,\mathrm{d}x_{n-2}\ldots\mathrm{d}x_2\,\mathrm{d}x_1 \\ &= \frac{n-1}{n!}. \end{aligned}$$

This formula can be also derived without integration using the following consideration. All $n!$ orderings of variables $X_1, \ldots, X_n$ are equally probable. The event

$$X_1 > X_2 > \cdots > X_{n-2} > X_{n-1} < X_n$$

occurs only in the following $n-1$ cases:

$$\begin{array}{l} X_n > X_1 > X_2 > \cdots > X_{n-2} > X_{n-1}, \\ X_1 > X_n > X_2 > \cdots > X_{n-2} > X_{n-1}, \\ X_1 > X_2 > X_n > \cdots > X_{n-2} > X_{n-1}, \\ \cdots\cdots\cdots\cdots\cdots\cdots\cdots\cdots\cdots\cdots \\ X_1 > X_2 > X_3 > \cdots > X_n \quad > X_{n-1}. \end{array}$$

Hence from the classical definition of probability we also get $C_n = (n-1)/n!$.

Let A_n be the probability that an American catches exactly n fish. Define $S_n = X_1 + \cdots + X_n$. It is known (see Cramér 1946, Sect. 19.1) that S_n has density

$$f_n(s) = \frac{1}{(n-1)!} \sum_{i=0}^{\lfloor s \rfloor} (-1)^i \binom{n}{i} (s-i)^{n-1}, \qquad 0 < s < n.$$

In particular

$$f_n(s) = \frac{s^{n-1}}{(n-1)!} \quad \text{for} \quad 0 < s < 1.$$

Since the variables S_{n-1} and X_n are independent, their joint density equals the product of marginal densities. Therefore

$$A_n = \mathsf{P}(S_{n-1} < 1, X_n > 1 - S_{n-1}) = \int_0^1 \left(\int_{1-s}^1 \mathrm{d}x_n \right) \frac{s^{n-2}}{(n-2)!} \, \mathrm{d}s = \frac{n-1}{n!}.$$

Probabilities C_n and A_n are equal, so that from this point of view the rules of both countries allow their citizens to catch the same expected number of fish. However, it can be proved that the expected weights of captive fish (fish that have been caught) are not equal.

It is not difficult to verify that C_n are probabilities — that their sum equals to 1. We have $C_1 = 0$, and complete induction yields

$$C_2 + \cdots + C_n = 1 - \frac{1}{n!}.$$

The expected number of captive fish is

$$\sum_{n=2}^{\infty} n C_n = \sum_{n=2}^{\infty} \frac{1}{(n-2)!} = e.$$

7.5 BIRDS

There is a wire connecting two columns. On the wire n *birds* land at random. Each bird watches its nearest neighbor. What is the expected number of unwatched birds?

We first specify the problem. Assume that the length of the wire is a. Let $X_1 \le \cdots \le X_n$ be the points where birds landed. Assume that $X_1, \ldots, X_n$ is

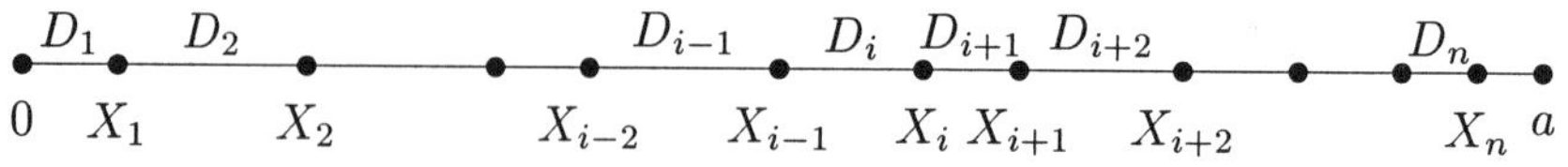

$$\overset{D_1}{\bullet} \quad \overset{D_2}{\bullet} \quad \bullet \quad \overset{D_{i-1}}{\bullet} \quad \overset{D_i}{\bullet} \quad \overset{D_{i+1}}{\bullet} \quad \overset{D_{i+2}}{\bullet} \quad \bullet \quad \overset{D_n}{\bullet} \quad \bullet$$
$$0 \quad X_1 \quad X_2 \quad X_{i-2} \quad X_{i-1} \quad X_i \; X_{i+1} \quad X_{i+2} \quad X_n \quad a$$

Fig. 7.1 Positions of birds.

a sample from the rectangular distribution $\mathsf{R}(0, a)$. The joint density of random variables $X_1, \ldots, X_n$ is

$$f(x_1, \ldots, x_n) = \begin{cases} n!/a^n & \text{for} \quad 0 < x_1 < \cdots < x_n < a, \\ 0 & \text{otherwise} \end{cases}$$

(see Hájek and Šidák 1968, p. 38). Define

$$D_1 = X_1, \qquad D_i = X_i - X_{i-1} \text{ for } i = 2, \ldots, n.$$

The inverse transformation is

$$X_i = D_1 + \cdots + D_i$$

and its Jacobian is 1. Hence the joint density of variables $D_1, \ldots, D_n$ is

$$g(d_1, \ldots, d_n) = \begin{cases} n!/a^n & \text{for } 0 < d_1 < a, \ldots, 0 < d_n < a, \; \sum_{i=1}^n d_i < a, \\ 0 & \text{otherwise.} \end{cases}$$

Density g is a symmetric function, and so variables $D_1, \ldots, D_n$ are called *exchangeable* (see Rényi 1970, p. 156) or *symmetric* (see Štěpán 1987, p. 399).

Assume first that $n \geq 4$. There are three kinds of birds:

1. Two birds at both ends of the line
2. Two birds in penultimate positions
3. Remaining $n - 4$ birds

The bird at the left end is unwatched in the case $D_3 < D_2$. Because of exchangeability, $\mathsf{P}(D_3 < D_2) = \mathsf{P}(D_2 < D_3) = \frac{1}{2}$. The next bird, its neighbor, is always watched, since the end bird watches it. If $3 \leq i \leq n - 2$, then the ith bird is unwatched when $D_{i-1} < D_i$ and $D_{i+2} < D_{i+1}$ (see Fig. 7.1). Using exchangeability, we find that this probability is $\frac{1}{4}$. The situation at the right end of the wire is similar to that at the left end. A randomly chosen bird occupies position i with probability $1/n$. Using the theorem of total probability, we find that a randomly chosen bird is unwatched with probability

$$P_n = \frac{1}{n} \times \frac{1}{2} + \frac{1}{n} \times 0 + (n-4) \times \frac{1}{n} \times \frac{1}{4} + \frac{1}{n} \times 0 + \frac{1}{n} \times \frac{1}{2} = \frac{1}{4}.$$

It is interesting that for $n \geq 4$ this probability does not depend on n.

Let

$$Y_i = \begin{cases} 1 & \text{if the } i\text{th bird is unwatched,} \\ 0 & \text{otherwise.} \end{cases}$$

Define

$$p_i = \mathsf{P}(Y_i = 1).$$

We already calculated that

$$p_1 = p_n = \frac{1}{2}, \qquad p_2 = p_{n-1} = 0, \qquad p_i = \frac{1}{4} \text{ for } 3 \le i \le n-2.$$

The number of unwatched birds is $Y = Y_1 + \cdots + Y_n$. Its expectation is

$$\mathsf{E}Y = \sum_{i=1}^{n} \mathsf{E}Y_i = 2 \times \frac{1}{2} + (n-4) \times \frac{1}{4} = \frac{n}{4}.$$

Let e_i be expected number of birds that are watched by i other birds: $i = 0, 1, 2$. The total number of birds is n, and thus we have $e_0 + e_1 + e_2 = n$. The number of birds watched by no bird, one bird, or two birds, is also n, and so we have $0 \times e_0 + 1 \times e_1 + 2 \times e_2 = n$. Hence $e_2 = n/4$.

If $n = 3$, we have two birds at both ends of the wire and only one penultimate bird. The probability that a randomly chosen bird is unwatched is

$$P_3 = \frac{1}{3} \times \frac{1}{2} + \frac{1}{3} \times 0 + \frac{1}{3} \times \frac{1}{2} = \frac{1}{3}.$$

The expected number of unwatched birds is 1, since exactly one of the three birds is unwatched and exactly one bird (the penultimate one) is watched by both remaining birds.

The case $n = 2$ is trivial.

Solution of this problem was published in *Am. Math. Monthly* (**89**, 1982, pp. 274–275).

7.6 SULTAN AND CALIPH

"If two of my children are selected at random, likely as not they will be of the same sex", said the *sultan* to the *caliph*.

"What are the chances that both will be girls?" asked the caliph.

"Equal to the chance that one child selected at random will be a boy," replied the sultan. How many children did the sultan have?

Rabinowitz (1992, p. 290), mentions that this problem was published in *Parabola* (**19**, 1983/2, p. 23), with its solution in the same journal (**20**, 1984/1, p. 33).

Assume that the sultan has a boys and b girls. Let $a > 0$, $b > 0$. From these children $\binom{a+b}{2}$ pairs can be formed. Obviously,

$$\mathsf{P}(\text{two boys}) = \frac{\binom{a}{2}}{\binom{a+b}{2}}, \qquad \mathsf{P}(\text{two girls}) = \frac{\binom{b}{2}}{\binom{a+b}{2}},$$

$$\mathsf{P}(\text{boy and girl}) = \frac{ab}{\binom{a+b}{2}}, \qquad \mathsf{P}(\text{boy}) = \frac{a}{a+b}.$$

From the sultan's remarks we know that

$$\frac{2ab}{(a+b)(a+b-1)} = \frac{a(a-1)+b(b-1)}{(a+b)(a+b-1)},$$

$$\frac{b(b-1)}{(a+b)(a+b-1)} = \frac{a}{a+b}.$$

This can also be written in the form

$$a^2 - 2ab + b^2 - a - b = 0,$$
$$a^2 + ab - b^2 - a + b = 0.$$

Adding both equations, we get $b = 2a - 2$. Inserting terms into any of the two last equations, we obtain the quadratic equation $a^2 - 7a + 6 = 0$. Its roots are $a_1 = 6$, $a_2 = 1$. Corresponding numbers of girls are $b_1 = 10$, $b_2 = 0$. Only a_1 and b_1 satisfy all conditions of the problem. Consequently, the sultan has 16 children: 6 boys and 10 girls.

7.7 PENALTIES

Ice hockey teams A and B shoot *penalties*. One round means that A shoots a penalty and B shoots a penalty. In a given round, if one team scores and the other does not, it is the end of the game and the successful team wins. Otherwise both teams continue to play another round. Assume that teams A and B score with probabilities $a \in (0,1)$ and $b \in (0,1)$, respectively, and that penalties can be considered as independent trials. Find probabilities $\mathsf{P}(A)$ and $\mathsf{P}(B)$ that A and B win, respectively. Calculate the expected number of rounds needed for decision.

The probability that team A wins in the first round is $p = a(1-b)$. The probability that team B wins in the first round is $q = (1-a)b$. The probability that the first round is indecisive is $r = ab + (1-a)(1-b)$. Of course, $p + q + r = 1$. Probabilities in the further rounds are the same, if the end does not come sooner. Probabilities that A and B win in the nth round are $P_n(A) = r^{n-1}p$ and $P_n(B) = r^{n-1}q$, respectively. Hence

$$\mathsf{P}(A) = \sum_{n=1}^{\infty} P_n(A) = \frac{p}{1-r}, \qquad \mathsf{P}(B) = \sum_{n=1}^{\infty} P_n(B) = \frac{q}{1-r}.$$

The probability p_n that nth round decides is

$$p_n = P_n(A) + P_n(B) = r^{n-1}(p+q) = r^{n-1}(1-r).$$

The expected number of rounds needed for decision is

$$\sum_{n=1}^{\infty} np_n = (1-r)\sum_{n=1}^{\infty} nr^{n-1} = \frac{1}{1-r}.$$

7.8 TWO 6S AND TWO 5S

Assume that we throw n *dice* once. Assume that $n \geq 4$. Find the probability that we obtain exactly two 6s and two 5s. Calculate n that maximizes P_n.

We have 6^n cases. Calculate the favorable cases. Two dice with 6s can be chosen in $\binom{n}{2}$ ways. Two dice with 5s among the remaining dice can be chosen in $\binom{n-2}{2}$ ways. Other dice, which must not show 5s or 6s, can be thrown in 4^{n-4} ways. Combining all the possibilities, we get

$$P_n = \frac{\binom{n}{2}\binom{n-2}{2}4^{n-4}}{6^n} = \frac{n(n-1)(n-2)(n-3)}{4^5}\left(\frac{2}{3}\right)^n.$$

Readers familiar with probability theory would write P_n without hesitation as the probability in the *trinomial distribution.* Since

$$\frac{P_{n+1}}{P_n} = \frac{2n+2}{3n-9},$$

sequence $\{P_n\}$ increases for $4 \leq n \leq 11$ and decreases for $n \geq 12$. Its maximal value is $P_{11} = P_{12} = 0.089$.

There are many similar problems. Rabinowitz (1992, p. 287), writes that the following problem was published in *The College Math. J.* (**15**, 1984, p. 347), with its solution in the same journal (**17**, 1986, p. 251). Determine m, the number of dice required to maximize the probability of obtaining exactly n 6s and $m - n$ 5s when the dice are thrown once.

We have 6^n cases, and $\binom{n}{m}$ of them are favorable. Hence

$$P_n^* = \frac{\binom{n}{m}}{6^n}.$$

Since

$$\frac{P_{n+1}^*}{P_n^*} = \frac{n+1}{6n-6m+6},$$

sequence $\{P_n^*\}$ is increasing for $m \leq n < 6m/5 - 1$ and decreasing for $n > 6m/5 - 1$.

Rabinowitz (1992, p. 287), mentions another problem that was introduced in *The Two Year College Math. J.* (**12**, 1981, p. 275), with its solution in the same journal (**14**, 1983, p. 71). Determine the number of dice required to maximize the probability of obtaining exactly n 6s when the dice are thrown once.

We have 6^n cases, and $\binom{n}{m}5^{n-m}$ of them are favorable. The probability of obtaining exactly n 6s is

$$P'_n = \frac{\binom{n}{m}5^{n-m}}{6^n} = \frac{\binom{n}{m}}{5^m}\left(\frac{5}{6}\right)^n.$$

Since

$$\frac{P'_{n+1}}{P'_n} = \frac{5n+5}{6n-6m+6},$$

sequence $\{P'_n\}$ is increasing for $m \le n < 6m-1$ and decreasing for $n > 6m-1$. We have $P'_{6m-1} = P'_{6m}$. For example, if we want to obtain exactly one 6, we have maximal probability when we roll 5 or 6 dice.

7.9 PRINCIPLE OF INCLUSION AND EXCLUSION

We have n *dice*. The dice are thrown once. Assume that $n \ge 6$ and find the probability P_n that each number from 1 to 6 will be represented at least once. Calculate n such that the difference between P_n and 0.5 is smallest. Find the smallest number n such that $P_n \ge 0.95$.

Problems of this kind are seen quite frequently. Rabinowitz (1992, p. 287), writes that one of them was published in *J. Recreational Math.* (**12**, 1980, p. 219), with its solution in the same journal (**13**, 1981, p. 225).

If $n = 6$, then calculation is simple. We have 6^6 possible results, but all numbers from 1 to 6 are represented in only 6! of them. Thus we have $P_6 = 6!/6^6 = 0.015$.

If $n > 6$, then we must use another approach. We have 6^n possible results. There are $\binom{6}{1}5^6$ results such that at least one number is missing. Similarly, there are $\binom{6}{2}4^6$ results such that at least two numbers are missing. However, the difference $6^n - \binom{6}{1}5^6$ does not represent the number of results, where all numbers are represented, because some cases (e.g., if 1 and 2 are missing simultaneously) were subtracted twice. We must apply some corrections, which is a procedure known from set theory. In this way we find that the number of results such that all numbers are represented is

$$6^n - \binom{6}{1}5^n + \binom{6}{2}4^n - \binom{6}{3}3^n + \binom{6}{4}2^n - \binom{6}{5}1^n.$$

This method of calculation is called the *principle of inclusion and exclusion*. A detailed explanation can be found, for example, in Feller (1968, Vol. I, Chap. IV).

Now, it is easy to see that

$$P_n = 1 - 6\left(\frac{5}{6}\right)^n + 15\left(\frac{4}{6}\right)^n - 20\left(\frac{3}{6}\right)^n + 15\left(\frac{2}{6}\right)^n - 6\left(\frac{1}{6}\right)^n.$$

Of course, probability P_6 is the same as that calculated at the beginning. It can be verified numerically that $P_{13} = 0.514$ is the value nearest to 0.5 and that $P_{27} = 0.957$ is the value with the smallest index such that it is equal to or greater than 0.95.

7.10 MORE HEADS ON COINS

Player A tosses n *coins*, and player B tosses $n + r$ coins. Assume that $r \geq 1$. Find the probability $P_{n,r}$ that B gets more heads than A.

The probability that A gets exactly i heads is $\binom{n}{i}2^{-n}$. The probability that B gets exactly j heads is $\binom{n+r}{j}2^{-n-r}$. The probability that B gets k heads more than A is

$$Q_k = 2^{-2n-r}\sum_{i=0}^{n+r-k}\binom{n}{i}\binom{n+r}{i+k} = 2^{-2n-r}\sum_{i=0}^{n+r-k}\binom{n}{i}\binom{n+r}{n+r-k-i}.$$

The *additive theorem for binomial coefficients* (which is also known as *Cauchy combinatorial form*) gives

$$\sum_{\nu=0}^{m}\binom{x}{\nu}\binom{y}{m-\nu} = \binom{x+y}{m}$$

(see Kaucký 1975, p. 45). The additive theorem can be proved using complete induction with respect to m. Therefore, we have

$$Q_k = 2^{-2n-r}\binom{2n+r}{n+r-k}.$$

This implies that

$$P_{n,r} = \sum_{k=1}^{n+r} Q_k = 2^{-2n-r}\sum_{k=1}^{n+r}\binom{2n+r}{n+r-k} = 2^{-2n-r}\sum_{k=1}^{n+r}\binom{2n+r}{n+k}.$$

Since

$$\sum_{k=1}^{n+r}\binom{2n+r}{n+k} = \sum_{h=0}^{n+r-1}\binom{2n+r}{2n+r-h} = \sum_{h=0}^{n+r-1}\binom{2n+r}{h},$$

we have

$$\begin{aligned} P_{n,r} &= 2^{-2n-r-1}\left[\sum_{h=0}^{n+r-1}\binom{2n+r}{h} + \sum_{h=n+1}^{2n+r}\binom{2n+r}{h}\right] \\ &= 2^{-2n-r-1}\left[\sum_{h=0}^{2n+r}\binom{2n+r}{h} + \sum_{h=n+1}^{n+r-1}\binom{2n+r}{h}\right] \\ &= \frac{1}{2} + 2^{-2n-r-1}\sum_{h=n+1}^{n+r-1}\binom{2n+r}{h}. \end{aligned}$$

In particular, we obtain

$$\begin{aligned} P_{n,1} &= \frac{1}{2}, \\ P_{n,2} &= \frac{1}{2}\left[1 + \frac{1}{2^{2n+2}}\binom{2n+2}{n+1}\right]. \end{aligned}$$

Rabinowitz (1992, p. 286), remarks that for $r = 1$, this problem was formulated in *Function* (**6**, 1982/2, p. 26), with its solution in the same journal (**6**, 1982/4, p. 27). The same problem was also posed in the Australian Mathematical Olympiad. For $n = 9$ and $r = 2$, the problem was presented in *Crux Mathematicorum* (**6**, 1980, p. 109), with its solution in the same journal (**6**, 1980, p. 149 and p. 311).

7.11 HOW COMBINATORIAL IDENTITIES ARE BORN

A gambling student tosses a fair *coin*. He scores one point for each head that turns up and two points for each tail. Find probability P_n that the student scores exactly n points sometimes during this experiment. Rabinowitz (1992, p. 286), mentions that this problem was published in *Parabola* (**17**, 1981/1, p. 25), with its solution in the same journal (**17**, 1981/3, p. 26). The problem was also posed in the Canadian Mathematical Olympiad.

The student scores exactly n points if and only if one of the following disjoint events occurs:

- In the first n tosses he gets n heads and no tail
- In the first $n-1$ tosses he gets $n-2$ heads and 1 tail
- In the first $n-2$ tosses he gets $n-4$ heads and 2 tails
- And so on

Let $\lfloor a \rfloor$ be the integer part of the number a. Then the probability P_n can be written in the form

$$P_n = \sum_{i=0}^{\lfloor \frac{n}{2} \rfloor} \binom{n-i}{n-2i} \frac{1}{2^{n-i}} = 2^{-n} \sum_{i=0}^{\lfloor \frac{n}{2} \rfloor} 2^i \binom{n-i}{i}.$$

However, the author does not know how to transform this formula into a simpler form. Thus we try to calculate P_n by another method. In order to score n points, the student must score either in the first toss 1 point and then $n-1$ points, or in the first toss 2 points and then $n-2$ points. The theorem of total probability gives the homogeneous difference equation

$$P_n = \frac{1}{2} P_{n-1} + \frac{1}{2} P_{n-2}, \qquad n \geq 2.$$

We have initial conditions $P_0 = 1$, $P_1 = \frac{1}{2}$. The characteristic equation

$$x^2 - \frac{1}{2}x - \frac{1}{2} = 0$$

has roots $x_1 = 1$, $x_2 = -\frac{1}{2}$. Then a general solution of the difference equation is

$$P_n = A + B\left(-\frac{1}{2}\right)^n.$$

From the initial conditions we get $A = \frac{2}{3}$, $B = \frac{1}{3}$. Finally, we can write

$$P_n = \frac{2 + \left(-\frac{1}{2}\right)^n}{3}, \qquad n \geq 0.$$

Comparing both formulas for P_n, we obtain a combinatorial identity

$$2^{-n}\sum_{i=0}^{\lfloor \frac{n}{2} \rfloor} 2^i \binom{n-i}{i} = \frac{2 + \left(-\frac{1}{2}\right)^n}{3}.$$

The formula for P_n shows that $P_n \to \frac{2}{3}$ as $n \to \infty$.

7.12 EXAMS

A teacher knows there is a probability of $p \in (0, 1)$ that a student in her class taking an examination may know nothing about the subject.

The teacher asks n independent questions. If the student understands the subject, he answers with probability u any single question correctly. But there is also a probability $b \in (0, 1)$ that a student who does not know the subject can simply guess the correct answer. Assume that $u > b$. The teacher found out that the student answered k questions correctly. Find the probability $P_k(n)$ that the student knows the subject. This problem is introduced in Stirzaker (1994, p. 47, Problem 17).

Let A be the event that the student understands the subject, and N the event that he does not understand it. Let K be the event that the student correctly answers k questions. Define $q = 1 - p$. Bayes' theorem yields

$$\begin{aligned} P_k(n) &= \mathsf{P}(A|K) = \frac{\mathsf{P}(K|A)\mathsf{P}(A)}{\mathsf{P}(K|A)\mathsf{P}(A) + \mathsf{P}(K|N)\mathsf{P}(N)} \\ &= \frac{\binom{n}{k} u^k (1-u)^{n-k} q}{\binom{n}{k} u^k (1-u)^{n-k} q + \binom{n}{k} b^k (1-b)^{n-k} p} \\ &= \frac{q u^k (1-u)^{n-k}}{q u^k (1-u)^{n-k} + p b^k (1-b)^{n-k}}. \end{aligned}$$

If the student answers all questions correctly, we have $k = n$ and

$$P_n(n) = \frac{qu^n}{qu^n + pb^n} = \frac{1}{1 + \dfrac{p}{q}\left(\dfrac{b}{u}\right)^n}.$$

Since we assume that $u > b$, we can see that $P_n(n) \to 1$ as $n \to \infty$.

We can also ask how many questions the student must answer correctly to convince the teacher that it is more likely he understands the subject, than does not. We have already calculated $\mathsf{P}(A|K)$. We find quite analogously that

$$\mathsf{P}(N|K) = \frac{pb^k(1-b)^{n-k}}{qu^k(1-u)^{n-k} + pb^k(1-b)^{n-k}}.$$

Inequality $\mathsf{P}(A|K) > \mathsf{P}(N|K)$ holds if and only if k fulfills the condition

$$\left[\frac{u(1-b)}{b(1-u)}\right]^k > \frac{p}{q}\left(\frac{1-b}{1-u}\right)^n.$$

Would the student convince the teacher that he knows the subject even if he correctly answers all questions? In this case $n = k$, and the formula reads

$$\left(\frac{u}{b}\right)^n > \frac{p}{q}.$$

For $u > n$, the teacher can easily determine n such that the condition is fulfilled.

7.13 WYVERNS

It is known that *wyverns* (mythical animals, e.g., unicorns) which are frequently depicted on coats-of arms, belong to very endangered animals in the wild. A person wishes to form a captive breeding colony. It is estimated that a viable colony should initially contain r males and r females. A male is trapped with probability $p \in (0, 1)$ and a female with probability $q = 1 - p$. Trappings of individual animals are independent events. Find the probability p_n that it is necessary to capture n animals in order to have r males and r females (see Stirzaker 1994, p. 56).

Let A_n be the event that we have r males and r females with the nth capture. Let M and F be the events that the nth captured animal is male or female, respectively. The event $A_n \cap M$ occurs in the case that there are $r - 1$ males and $n - r$ females among the first $n - 1$ captured animals and that the next captured animal is male. Hence

$$\mathsf{P}(A_n \cap M) = \binom{n-1}{r-1} p^{r-1} q^{n-r} p.$$

Similarly, we obtain

$$\mathsf{P}(A_n \cap F) = \binom{n-1}{r-1} p^{n-r} q^{r-1} q.$$

This implies that for $n \geq 2r$ we have

$$p_n = \mathsf{P}(A_n) = \mathsf{P}(A_n \cap M) + \mathsf{P}(A_n \cap F) = \binom{n-1}{r-1} p^r q^r (p^{n-2r} + q^{n-2r}).$$

Let X be a random variable equal to the number of captured animals needed for having r males and r females. It is clear that

$$\mathsf{E}X = \sum_{n=2r}^{\infty} np_n = \left(\frac{p}{q}\right)^r \sum_{n=2r}^{\infty} n\binom{n-1}{r-1} q^n + \left(\frac{q}{p}\right)^r \sum_{n=2r}^{\infty} n\binom{n-1}{r-1} p^n.$$

Since $n\binom{n-1}{r-1} = r\binom{n}{r}$, we have

$$\mathsf{E}X = r\left(\frac{p}{q}\right)^r \sum_{n=2r}^{\infty} \binom{n}{r} q^n + r\left(\frac{q}{p}\right)^r \sum_{n=2r}^{\infty} \binom{n}{r} p^n.$$

Formulas (5.7) and (5.9) imply that

$$\begin{aligned} 1 &= \sum_{k=0}^{\infty} \binom{r+k-1}{k} p^r q^k = \frac{p^r}{q^{r-1}} \sum_{k=0}^{\infty} \binom{r+k-1}{r-1} q^{r+k-1} \\ &= \frac{p^r}{q^{r-1}} \sum_{n=r-1}^{\infty} \binom{n}{r-1} q^n. \end{aligned}$$

Thus we can see that for all integers $h \geq 0$ we have

$$\sum_{n=h}^{\infty} \binom{n}{h} q^n = \frac{q^h}{p^{h+1}}.$$

After some arrangement we obtain

$$\mathsf{E}X = \frac{r}{pq} - p^r \sum_{n=r}^{2r-1} \binom{n}{r} q^{n-r} - q^r \sum_{n=r}^{2r-1} \binom{n}{r} p^{n-r}.$$

Expectation $\mathsf{E}X$ is introduced in Table 7.2 for some values r and p.

7.14 GAPS AMONG BALLS

Cells are located on the real line at points $0, 1, 2, \ldots$. In each of n independent trials a *ball* is placed in the cell i with probability $2^{-(i+1)}$, $i = 0, 1, 2, \ldots$. Let K be the number of empty cells located anywhere before the occupied cell with the largest number. We briefly state that the balls create K gaps. Prove that we have for all $n \geq 1$

$$\mathsf{P}_n(K = k) = 2^{-(k+1)}, \qquad k = 0, 1, 2, \ldots.$$

Table 7.2 Expected number of captured animals.

r	$p = 0.5$	$p = 0.75$
2	6.75	9.48
5	18.49	25.36
10	38.35	52.00
20	78.25	105.33
50	198.16	265.33
100	398.11	532.00

A solution of this problem was published by D. E. Knuth, the author of TEX, in *Am. Math. Monthly* (**94**, 1987, p. 189).

Each trial can be equivalently expressed in such a way that we place a ball with probability $\frac{1}{2}$ into the cell 0. If the trial fails, the process is repeated with the next cell. Knuth's solution is based on a trial in which we try to place n balls into cell 0. If we do not succeed with j of them, it is necessary to try to place them into cell 1. We investigate how the event $\{K = k\}$ can occur.

All balls are placed into cell 0 with probability 2^{-n}. This case corresponds only to the event $\{K = 0\}$.

Otherwise $n-j$ balls are placed into cell 0 with probability $\binom{n}{j}2^{-n}$. The remaining j balls must create k gaps (but we consider now only cases $j = 1, 2, \ldots, n-1$).

With probability 2^{-n} no ball is placed in the cell 0. It remains empty. If k gaps are to arise, n balls must create only $k-1$ gaps. This case does not correspond to the event $\{K = 0\}$.

Define $\mathsf{P}_n(K = k) = 0$ for $k < 0$ and

$$\delta_{ij} = \begin{cases} 1 & \text{for } i = j, \\ 0 & \text{for } i \neq j. \end{cases}$$

Our analysis leads to

$$\mathsf{P}_n(K = k) = 2^{-n}\delta_{k0} + 2^{-n}\sum_{j=1}^{n-1}\binom{n}{j}\mathsf{P}_j(K = k) + 2^{-n}\mathsf{P}_n(K = k-1).$$

In particular, we get

$$\mathsf{P}_n(K = 0) = 2^{-n} + 2^{-n}\sum_{j=1}^{n-1}\binom{n}{j}\mathsf{P}_j(K = 0).$$

It is clear that

$$\mathsf{P}_1(K = k) = 2^{-(k+1)}, \qquad k = 0, 1, 2, \ldots.$$

We can prove by complete induction that

$$\mathsf{P}_n(K = 0) = 2^{-1}.$$

We start with the proof of formula for $\mathsf{P}_n(K = k)$ when $k \geq 1$. We again use complete induction. From the recurrent formula, which was already derived for $\mathsf{P}_n(K = k)$, we obtain

$$\mathsf{P}_n(K = k) = 2^{-n} \sum_{j=1}^{n-1} \binom{n}{j} \mathsf{P}_j(K = k) + 2^{-n}\mathsf{P}_n(K = k - 1).$$

If we assume that formula $\mathsf{P}_j(K = k) = 2^{-(j+1)}$ holds for all k if $j \leq n - 1$ and for values smaller than k if $j = n$, we have

$$\mathsf{P}_n(K = k) = 2^{-n} \sum_{j=1}^{n-1} \binom{n}{j} 2^{-(k+1)} + 2^{-n}2^{-k} = 2^{-(k+1)}.$$

This concludes the proof.

7.15 NUMBERED PEGS

A series of holes for *pegs* is in a long narrow board. Holes and pegs are numbered by numbers $1, 2, \ldots, n$. Pegs are randomly chosen in succession and plugged into the corresponding hole. Find the probability that the holes will be filled continuously. This means that if the peg with the number k is chosen first, the next peg must have either $k - 1$, or $k + 1$. If pegs $p, p + 1, p + 2, \ldots, q$ are already chosen, the next peg must have either $p - 1$ or $q + 1$. Rabinowitz (1992, p. 289), indicates that this problem was published in *The Pi Mu Epsilon Journal* (**7**, 1984, p. 671), with its solution in the same journal (**8**, 1985, p. 138).

Let $p_n(k)$ be the conditional probability that n pegs will fill in holes continuously, given that the first peg has the number k. It is clear that

$$p_n(1) = \frac{1}{(n-1)!}, \qquad p_n(n) = \frac{1}{(n-1)!}.$$

The kernel of our next consideration is that after choice of pegs k and $k - 1$ our situation is the same as if we started with only $n - 1$ pegs and the peg $k - 1$ was chosen first. Similarly, after choice of pegs k and $k + 1$ the situation is the same as if we started only with $n - 1$ pegs and peg k was chosen first. The theorem of total probability implies that

$$p_n(k) = \frac{1}{n-1} p_{n-1}(k - 1) + \frac{1}{n-1} p_{n-1}(k), \qquad 2 \leq k \leq n - 1.$$

It can be proved by complete induction that

$$p_n(k) = \frac{1}{(k-1)!\,(n-k)!}.$$

This formula is valid not only for $2 \leq k \leq n-1$ but also for $k = 1$ and $k = n$. The next application of theorem of total probability gives the result

$$p_n = \frac{1}{n}\sum_{k=1}^{n} p_n(k) = \frac{1}{n}\frac{1}{(n-1)!}\sum_{k=1}^{n}\binom{n-1}{k-1} = \frac{2^{n-1}}{n!}.$$

7.16 CRUX MATHEMATICORUM

The journal *Crux Mathematicorum* publishes also problems from probability theory. Rabinowitz (1992, p. 294), writes that the next problem was published in this journal (**7**, 1981, p. 19), with its solution in the same journal (**8**, 1982, p. 51).

Four balls marked with letters C, R, U, and X are placed in a urn. We draw n balls in succession with replacement. Let P_n be the probability that the term CRUX appears in any four consecutive extractions.

1. Calculate the minimum n for which $P_n > 0.99$.

2. Find an explicit formula for P_n as a function of n.

It is easy to find out that $P_n = 0$ for $n \leq 3$ and that

$$P_4 = \frac{1}{4^4}, \qquad P_5 = \frac{2\times 4}{4^5}, \qquad P_6 = \frac{3\times 4^2}{4^6}, \qquad P_7 = \frac{4\times 4^3}{4^7},$$

$$P_8 = \frac{5\times 4^4 - 1}{4^8}, \quad P_9 = \frac{6\times 4^5 - 3\times 4}{4^9}, \quad \ldots$$

Let $n \geq 5$. Then CRUX can appear only in the following two disjoint cases:

1. CRUX is extracted already somewhere in the first $n-1$ draws.

2. CRUX is extracted for the first time as late as in the nth draw. Then CRUX was not extracted in the first $n-4$ draws (which has probability $1 - P_{n-4}$) and in the last four draws the letters C, R, U, and X appeared consecutively (which has probability 4^{-4}). We recall that all draws are independent.

This consideration leads to

$$P_n = P_{n-1} + \frac{1 - P_{n-4}}{4^4}, \qquad n \geq 5.$$

We have already introduced the initial values of P_n, and so it is easy to calculate probabilities P_n numerically, if n is not very large. Some results are introduced in Table 7.3.

Minimal n such that probability P_n is equal to or larger than 0.95 is 760 and $P_{760} = 0.950124$. Similarly, $n = 1166$ is minimal such that $P_n \geq 0.99$ and $P_{1166} = 0.990011$.

Table 7.3 Probability P_n.

n	P_n	n	P_n	n	P_n	n	P_n
10	0.027	300	0.692	700	0.937	1 000	0.981
50	0.170	400	0.792	800	0.957	1 100	0.987
100	0.319	500	0.860	900	0.971	1 200	0.991
200	0.542	600	0.906				

Table 7.4 Number of votes for DŽJ and for SPR—RSČ.

District	DŽJ	SPR—RSČ
Domažlice	1295	1295
Karlovy Vary	2664	2664
Uherské Hradiště	2105	2105

Let $\lfloor x \rfloor$ be an integer part of the number x. It can be proved by complete induction that

$$P_n = \sum_{i=1}^{\lfloor n/4 \rfloor} (-1)^{i+1} \binom{n-3i}{i} 4^{-4i}, \qquad n \geq 5.$$

The same formula can also be derived using the principle of inclusion and exclusion.

7.17 TIES IN ELECTIONS

Newspapers informed citizens that numbers of votes in some districts for certain political parties, especially for DŽJ (an unsuccessful party of Czech pensioners) and for SPR—RSČ (an unsuccessful Czech Republican Party, no longer represented in the Czech parliament) in *parliamentary elections* in the Czech Republic in June 1998 were equal (see Table 7.4). For example, the newspaper *Mladá Fronta Dnes* on June 29, 1998, in an article entitled "Results of elections in districts must be checked again" cited the Central Election Committee and introduced the following data: "During elections $5,969,505$ voters voted in 89 districts for 13 parties. Two independent Institutes of Academy of Sciences calculated for *Mladá Fronta Dnes* using very efficient computers that the probability of such a triple tie is somewhere from 1:10 000 to 1:6.7 million."

We denote the parties DŽJ and SPR—RSČ by A and B, respectively. Let a district have n voters. Let m be the integer part of number $n/2$. Let p_1 and p_2 be the probabilities that a randomly chosen voter votes for parties A and B, respectively. Then the voter votes with probability $1 - p_1 - p_2$ for one of the other parties or does not vote at all. Assume that voters vote independently.

The probability $P(i,j)$ that in this district parties A and B receive i and j votes, respectively, is given by *trinomic distribution* and is equal to

$$P(i,j) = \frac{n!}{i!j!(n-i-j)!} p_1^i p_2^j (1 - p_1 - p_2)^{n-i-j}.$$

The probability that A and B have the same number of votes is

$$P = \sum_{i=1}^{m} P(i,i).$$

In the next analysis we assume that $p_1 = p_2 = p$. Then

$$\begin{aligned} P &= \sum_{i=0}^{m} \frac{n!}{i!i!(n-2i)!} p^{2i} (1-2p)^{n-2i} \\ &= (1-2p)^n \sum_{i=0}^{m} \frac{n!}{i!i!(n-2i)!} \left[\frac{p^2}{(1-2p)^2} \right]^i . \end{aligned}$$

Define

$$z = \frac{p^2}{(1-2p)^2}, \qquad a_i = (1-2p)^n \frac{n!}{i!i!(n-2i)!} z^i.$$

It is clear that $a_i = P(i,i)$. For $1 \le i \le m$, we have

$$\frac{a_i}{a_{i-1}} = z \frac{(n-2i+2)(n-2i+1)}{i^2}.$$

Sequence $\{a_i\}$ increases until $a_i/a_{i-1} > 1$. This inequality is equivalent to

$$(1-4z)i^2 + 2z(2n+3)i - z(n+2)(n+1) < 0.$$

The quadratic equation

$$(1-4z)x^2 + 2z(2n+3)x - z(n+2)(n+1) = 0$$

has one positive root and one negative root. The positive root is

$$x_1 = \frac{-z(2n+3) + \sqrt{zn^2 + z(3n+2+z)}}{1-4z}.$$

In real elections n is very large and z is a very small positive number. To illustrate this point, we choose $n = 30,000$, $p = 0.04$, so that $m = 15,000$, $z = 0.00189036$. The number n was chosen in such a way that 1295 votes for SPR—RSČ in district Domažlice correspond roughly to 4% of the number of voters $n = 30,000$. Then we can use the approximation

$$x_1 \doteq n\sqrt{z} = n\frac{p}{1-2p} \doteq np.$$

In our numerical example we have $x_1 = 1200.06$, whereas the approximation makes $np = 1200$. Therefore, the sequence $\{a_i\}$ is increasing for $i \leq 1200$ and then decreases. Its maximal term is $a_{1200} = 0.000138256$. Values a_i are very small when the difference $|i - 1200|$ is large. For example, $a_{1100} = 1.38 \times 10^{-8}$, $a_{1300} = 1.86 \times 10^{-8}$. Taking the preceding values of parameters, computer calculations give $P = 0.0081$.

There are 89 districts. The probability, that parties A and B receive the same number of votes in at least 3 districts is

$$P_{3+} = 1 - \binom{89}{0} P^0 (1-P)^{89} - \binom{89}{1} P^1 (1-P)^{88} - \binom{89}{2} P^2 (1-P)^{87} = 0.036.$$

This is a small but not negligible, probability.

We simplified the derivation assuming that each district has $30,000$ voters. In fact, most districts had a larger number of voters and the numbers of voters were different. Numerical studies show that the final probability is a bit smaller, but not substantially. The order of the result is the same. The average number of voters in a district was $5,969,505 : 89 = 67,073$. For simplicity, assume that each district has $n = 60,000$ voters. Then the maximal probability a_i is $a_{2400} = 0.000069133$, and numerical calculation gives $P = 0.005758$. The probability P_{3+}, that parties A and B will have the same number of votes in at least three districts, is $P_{3+} = 0.015$.

7.18 CRAPS

Craps is a game with two dice. Everitt (1999) writes that the game originated in or near Mississippi about 1800 and was played by slaves at that time. There are many variants of this game. It seems that one of them, called the *pass line bet*, is most popular. We analyze it here.

A player rolls two *dice* and calculates a sum S of points on them. She wins if $S = 7$ or $S = 11$; she looses in cases $S = 2$ or $S = 3$ or $S = 12$. Any other result is called a *point*. After a point the player continues to roll dice until she obtains either sum $S = 7$ — and she looses, or the point (i.e., the same sum as in the first shoot) — and she wins. Calculate the probability P that the player wins.

Define $p_i = \mathsf{P}(S = i)$, $i = 2, \ldots, 12$. If X is the number of points on the first dice and Y number of points on the other dice, then the sum $S = X + Y$ is introduced in Table 7.5.

The number of cells in Table 7.5 is 36. The sum $S = 3$ occurs twice, namely, when $X = 1, Y = 2$ and $X = 2, Y = 1$. Thus $p_3 = \frac{2}{36}$. In a similar way we obtain all probabilities p_i introduced in Table 7.6

We mentioned that the player wins if she gets either sum 7 or 11 in the first shoot, which has probability $p_7 + p_{11}$. The player also wins if she has sum i in the first shoot, and in a following shoot she obtains the sum i before she shoots sum 7

Table 7.5 Sum of points on two dice.

	X					
Y	1	2	3	4	5	6
1	2	3	4	5	6	7
2	3	4	5	6	7	8
3	4	5	6	7	8	9
4	5	6	7	8	9	10
5	6	7	8	9	10	11
6	7	8	9	10	11	12

Table 7.6 Probabilities p_i.

i	2	3	4	5	6	7
p_i	$\frac{1}{36}$	$\frac{2}{36}$	$\frac{3}{36}$	$\frac{4}{36}$	$\frac{5}{36}$	$\frac{6}{36}$
i	8	9	10	11	12	
p_i	$\frac{5}{36}$	$\frac{4}{36}$	$\frac{3}{36}$	$\frac{2}{36}$	$\frac{1}{36}$	

$(i = 4, 5, 6, 8, 9, 10)$. This happens with probability

$$\sum_{k=1}^{\infty} p_i(1 - p_i - p_7)^k p_i = \frac{p_i^2}{p_i + p_7}.$$

Since the events are disjoint, complete probability P of the win is the sum

$$P = p_7 + p_{11} + \sum_{\substack{i=4 \\ i \neq 7}}^{10} \frac{p_i^2}{p_i + p_7}.$$

Inserting for p_i values from Table 7.6, we obtain $P = 0.492929$.

7.19 PROBLEM OF EXCEEDING 12

We throw a *dice* repeatedly and independently and sum the points consecutively. We stop throwing at the moment when the sum exceeds 12 for the first time. This means that the sum is equal to a number between 13 and 18. Calculate the probability that the sum is equal to the number n $(n = 13, \ldots, 18)$.

Before describing a solution of the problem, we derive an auxiliary result. Consider a particle that starts at point 0 on the real line and moves to the right by the number of units shown on the dice. Then the particle stops and waits for the next dice throw.

Let p_n be the probability that the particle in this walk stops at point n. Probabilities p_n can be calculated using the theorem of total probability. For example, if $n = 3$, we consider the problem in the following way. The particle stops at point 3 if we got 1 in the last throw and the particle was at point 2, or if we got 2 in the last throw and the particle was at point 1, or if we got 3 at the beginning of the play. This leads to the equation $p_3 = \frac{1}{6}p_2 + \frac{1}{6}p_1 + \frac{1}{6}$. Quite analogously we obtain also remainig equations of the system

$$\begin{aligned}
p_1 &= \frac{1}{6}, \\
p_2 &= \frac{1}{6}p_1 + \frac{1}{6}, \\
p_3 &= \frac{1}{6}p_2 + \frac{1}{6}p_1 + \frac{1}{6}, \\
p_4 &= \frac{1}{6}p_3 + \frac{1}{6}p_2 + \frac{1}{6}p_1 + \frac{1}{6}, \\
p_5 &= \frac{1}{6}p_4 + \frac{1}{6}p_3 + \frac{1}{6}p_2 + \frac{1}{6}p_1 + \frac{1}{6}, \\
p_6 &= \frac{1}{6}p_5 + \frac{1}{6}p_4 + \frac{1}{6}p_3 + \frac{1}{6}p_2 + \frac{1}{6}p_1 + \frac{1}{6}, \\
p_n &= \frac{1}{6}p_{n-1} + \frac{1}{6}p_{n-2} + \frac{1}{6}p_{n-3} + \frac{1}{6}p_{n-4} + \frac{1}{6}p_{n-5} + \frac{1}{6}p_{n-6}, \quad n \geq 7.
\end{aligned}$$

Now, we can calculate recurrently values introduced in Table 7.7. However, p_n is given in the form of a linear difference equation. The corresponding characteristic equation

$$x^6 - \frac{1}{6}x^5 - \frac{1}{6}x^4 - \frac{1}{6}x^3 - \frac{1}{6}x^2 - \frac{1}{6}x - \frac{1}{6} = 0$$

has the roots

$$x_1 = 1, \qquad x_2 = -0.670332,$$

$$x_{34} = -0.375695 \pm 0.570175\, i, \qquad x_{56} = 0.294195 \pm 0.668367\, i.$$

Thus the general solution of the difference equation is

$$p_n = \sum_{k=1}^{6} A_k x_k^n.$$

From the initial conditions, which specify values $p_1, \ldots, p_6$, we have

$$A_1 = 0.285714, \qquad A_2 = A_3 = A_4 = A_5 = A_6 = 0.142857.$$

Then it is obvious that

$$\lim_{n \to \infty} p_n = 0.285714$$

and the speed of convergence is demonstrated in values introduced in Table 7.7.

Table 7.7 Probabilities p_n.

n	p_n	n	p_n	n	p_n
1	0.166667	8	0.268094	15	0.286114
2	0.194444	9	0.280369	16	0.287071
3	0.226852	10	0.289288	17	0.286702
4	0.264660	11	0.293393	18	0.285587
5	0.308771	12	0.290830	19	0.284713
6	0.360232	13	0.279263	20	0.285621
7	0.253604	14	0.283540		

Now, we return to the original problem of exceeding 12. Let P_n be the probability that the final sum is n ($n = 13, \ldots, 18$). We stop with sum 13 in the case that in the last trial we got 1 and we had 12 before, or if we got 2 and we had 11 before, and so on. Similarly we consider individual cases in situations when we stop with sum 14, $\ldots$, 18. This gives

$$P_{13} = \tfrac{1}{6}p_{12} + \tfrac{1}{6}p_{11} + \tfrac{1}{6}p_{10} + \tfrac{1}{6}p_9 + \tfrac{1}{6}p_8 + \tfrac{1}{6}p_7 = 0.279263,$$

$$P_{14} = \tfrac{1}{6}p_{12} + \tfrac{1}{6}p_{11} + \tfrac{1}{6}p_{10} + \tfrac{1}{6}p_9 + \tfrac{1}{6}p_8 = 0.236996,$$

$$P_{15} = \tfrac{1}{6}p_{12} + \tfrac{1}{6}p_{11} + \tfrac{1}{6}p_{10} + \tfrac{1}{6}p_9 = 0.192313,$$

$$P_{16} = \tfrac{1}{6}p_{12} + \tfrac{1}{6}p_{11} + \tfrac{1}{6}p_{10} = 0.145585,$$

$$P_{17} = \tfrac{1}{6}p_{12} + \tfrac{1}{6}p_{11} = 0.145585,$$

$$P_{18} = \tfrac{1}{6}p_{12} = 0.048472.$$

Of course, $P_{13} + \cdots + P_{18} = 1$. We stop with the largest probability with a sum of points equal to 13.

8

Problems on calculating expectation

8.1 CHRISTMAS PARTY

Children were invited to a *Christmas party*. Each of them brought a present. Presents were collected in a large basket. All presents were different but identically wrapped. Going home, each child randomly selected a present from the basket. Find the expected number of children who carry home their own presents.

Assume that n children attended the party. We label the children by the numbers $1, 2, \ldots, n$. Let us introduce the random variables

$$X_i = \begin{cases} 1 & \text{if the } i\text{th child carries home his or her original present,} \\ 0 & \text{otherwise.} \end{cases}$$

The total number of children carrying home their original present is

$$S = X_1 + \cdots + X_n.$$

Presents can be ordered in $n!$ ways. There exist $(n-1)!$ orderings such that the present brought by the ith child is in the ith place. Thus

$$\mathsf{P}(X_i = 1) = \frac{(n-1)!}{n!} = \frac{1}{n}.$$

Further

$$\mathsf{E}X_i = 1 \times \mathsf{P}(X_i = 1) + 0 \times \mathsf{P}(X_i = 0) = \frac{1}{n}$$

and

$$\mathsf{E}S = \mathsf{E}X_1 + \cdots + \mathsf{E}X_n = 1.$$

Table 8.1 Probabilities p_i.

i	0	1	2	3	4	5	6
p_i	0.368	0.368	0.184	0.061	0.015	0.003	0.000

The expected number of children carrying home their original presents is 1 independent of n.

It is clear that

$$\mathsf{var}\, X_i = \mathsf{E}X_i^2 - (\mathsf{E}X_i)^2 = \frac{1}{n} - \frac{1}{n^2} = \frac{n-1}{n^2}.$$

Let $i \neq j$. Then

$$\mathsf{P}(X_iX_j = 1) = \mathsf{P}(X_i = 1, X_j = 1) = \frac{(n-2)!}{n!} = \frac{1}{n(n-1)},$$

so that $\mathsf{E}X_iX_j = 1/[n(n-1)]$. *Covariance* of variables X_i and X_j is

$$\mathsf{cov}\,(X_i, X_j) = \mathsf{E}X_iX_j - \mathsf{E}X_i\mathsf{E}X_j = \frac{1}{n(n-1)} - \frac{1}{n^2} = \frac{1}{n^2(n-1)}.$$

The *variance of sum* of random variables, which generally can be dependent, is given by the formula

$$\begin{aligned} \mathsf{var}\, S &= \mathsf{var} \sum_{i=1}^{n} X_i = \sum_{i=1}^{n} \mathsf{var}\, X_i + \sum\sum_{1 \le i \neq j \le n} \mathsf{cov}(X_i, X_j) \\ &= n\frac{n-1}{n^2} + n(n-1)\frac{1}{n^2(n-1)} = 1. \end{aligned}$$

Let $p_i(n)$ be the probability that exactly i children carry home their own presents. It can be proved that

$$p_i(n) = \frac{1}{i!}\sum_{k=0}^{n-i} \frac{(-1)^k}{k!}, \qquad i = 0, 1, \ldots, n$$

(see Feller 1968, Vol. I, Chap. IV, Sect. 4). Further, we have

$$p_i(n) \to p_i = \frac{1}{i!}e^{-1}$$

as $n \to \infty$. Probabilities p_i are introduced in Table 8.1. These are, in fact, probabilities of Poisson distribution with parameter $\lambda = 1$.

Convergence $p_i(n)$ to p_i is quite rapid. The first three decimals are identical for $p_i(n)$ as well as for p_i already when $n \geq 10$.

The *problem on confused secretary* is solved in the same way. She wrote n different letters and n different envelopes. However, she puts letters into envelopes

Table 8.2 Values E_n.

n	E_n	n	E_n	n	E_n	n	E_n
0	0	3	3	6	7.49	9	12.23
1	1	4	4.40	7	9.07	10	13.82
2	2	5	5.92	8	10.65	11	15.40

purely randomly. The number of letters inserted in the correct envelopes is the random variable S, which was already analyzed.

Our results can be used in solution of a *problem on tossing hats*. Rabinowitz (1992, p. 290) writes that this problem was published in *Ontario Secondary School Mathematics Bulletin* (**18**, 1982/1, p. 18), with its solution in the same journal (**18**, 1982/2, p. 9).

Eleven men toss their hats in the air, and the hats are picked up randomly. Each man who received his own hat leaves, and the remaining men toss the hats again. The process continues until every man has received his own hat again. How many rounds of tosses are expected?

Let E_n be the expected number of rounds when n men begin to toss their hats in the air. Probabilities $p_i(n)$ introduced above are probabilities that exactly i men receive their own hats in the first round. It is clear that $E_1 = 1$. Define $E_0 = 0$. Using the theorem of total expectation we get

$$E_n = 1 + p_0(n)E_n + p_1(n)E_{n-1} + \cdots + p_{n-1}(n)E_1 + p_n(n)E_0.$$

Values $E_3, E_4, \ldots$ can be calculated recurrently from this formula. Some of them are introduced in Table 8.2.

If eleven men toss their hats in the air, the expected number of rounds of this entertainment is 15.40.

8.2 SPAGHETTI

Consider a dish of *spaghetti*. Assume that the number of spaghetti strings is n. If n is a large number and the spaghetti are sufficiently long (a requirement that is usually satisfied), by looking at them we cannot determine which two ends belong to the same spaghetti string. Assume that we choose randomly and independently two ends of spaghetti on the dish and we stick them together. We repeat this process with the remaining strings. We again take two ends and stick them together. After some time no free ends remain, since we have created loops of different lengths. Find the expected number of the loops. It is also possible that we succeeded in juxtaposing the ends of the same spaghetti string, so that we got a loop of unit length. Find the expected number of loops of unit length.

We introduce a solution published in *Am. Math. Monthly* (**88**, 1981, p. 621, Problem E2831). Let L_n be the expected number of loops when there are n spaghetti with free ends in the dish. Let P_n be the probability that two randomly chosen ends belong to the same spaghetti. It is clear that

$$P_n = \frac{1}{2n-1}.$$

If we stick two ends, we obtain either a loop or a new spaghetti string — of course, a longer one. Therefore

$$L_n = (1 + L_{n-1})P_n + L_{n-1}(1 - P_n) = L_{n-1} + P_n = L_{n-1} + \frac{1}{2n-1}.$$

Since $L_1 = 1$, we have

$$L_n = 1 + \frac{1}{3} + \frac{1}{5} + \cdots + \frac{1}{2n-1}.$$

We could ask how L_n behaves asymptotically as $n \to \infty$. In this case we use the formula for L_n in the form

$$L_n = 1 + \frac{1}{2} + \frac{1}{3} + \cdots + \frac{1}{2n} - \frac{1}{2}\left(1 + \frac{1}{2} + \frac{1}{3} + \cdots + \frac{1}{n}\right).$$

Using the Euler formula (4.4) we obtain

$$L_n \doteq \ln 2n + \gamma - \frac{1}{2}(\ln n + \gamma) = \ln 2 + \frac{1}{2}\gamma + \frac{1}{2}\ln n \doteq 0.98175 + \frac{1}{2}\ln n = L_n^*.$$

To have a numerical illustration, we analyze the case when we have $n = 100$ spaghetti on the plate. We calculate that $L_{100} = 3.284342189\ldots$, whereas the approximation gives $L_{100}^* = 3.284340104\ldots$. The approximation is quite good. On the other hand, everybody is surprised by the small expected number L_{100} of loops. The function L_n grows very slowly, for example, $L_{1000} = 4.435632\ldots$.

Now, consider the other question. First, we label (at least theoretically) spaghetti by the numbers $1, 2, \ldots, n$. Define

$$X_i = \begin{cases} 1 & \text{if the } i\text{th spaghetti forms a loop of unit length,} \\ 0 & \text{otherwise.} \end{cases}$$

It is clear that total number of loops of unit length is $X = \sum X_i$. Since

$$\mathsf{P}(X_i = 1) = \frac{1}{2n-1},$$

we have

$$\mathsf{E}X = \sum_{i=1}^{n} \mathsf{E}X_i = n\mathsf{E}X_i = n\mathsf{P}(X_i = 1) = \frac{n}{2n-1}.$$

It is obvious that $\mathsf{E}X \to \frac{1}{2}$ as $n \to \infty$.

8.3 ELEVATOR

In a building p persons enter the *elevator* at the ground floor. There are n floors above the ground floor. The probability of each person exiting the elevator on any floor is the same, namely $1/n$. The persons act independently of one another. Determine the expected number of stops until the elevator is emptied.

This problem is very popular, and many authors write about it (Field 1978; Feller 1968, Example b.7 in Chap. I, Sect. 2, Problems 7–12 in Chap. II, Sect. 11, also Chap. IV, Sect. 2; *SIAM Review* **15**, 1973, pp. 793–796; Gaver and Powell 1971). The simplest approach is to introduce *indicator random variables*

$$Y_i = \begin{cases} 1 & \text{if the } i\text{th floor is a stop,} \\ 0 & \text{otherwise,} \end{cases}$$

where $i = 1, \ldots, n$. The total number of elevator stops is $Y = Y_1 + \cdots + Y_n$. The probability that no one gets off at floor i, is $(1 - 1/n)^p$. The probability that the elevator stops at floor i is

$$\mathsf{P}(Y_i = 1) = 1 - \left(1 - \frac{1}{n}\right)^p.$$

Since $\mathsf{E}Y_i = \mathsf{P}(Y_i = 1)$, the expected number of stops is

$$\mathsf{E}Y = n\left[1 - \left(1 - \frac{1}{n}\right)^p\right].$$

In many textbooks devoted to probability theory, for example, in the book by Feller cited above, one can find derivation of the formula

$$\pi_k = \mathsf{P}(Y = k) = \frac{1}{n^p}\left[\binom{n}{k}\sum_{j=0}^{k-1}(-1)^j\binom{k}{j}(k-j)^p\right], \qquad k = 1, \ldots, n.$$

To introduce a numerical example, assume that $n = 6$. Further, we use values $p = 3$, $p = 6$, and $p = 10$. Probabilities π_k and expectations $\mathsf{E}Y$ are introduced in Table 8.3.

An inverse problem is interesting from statistical point of view. Assume that the number n of floors above the ground floor is known and the number k of elevator stops is also known. The problem is to estimate the number p of persons who entered the elevator. However, we do not investigate this problem here.

Some results can be easily generalized. Let p_{ij} be the probability that the jth person stops at floor i, $(j = 1, \ldots, p;\ i = 1, \ldots, n)$. Let all persons act independently. Analogously to the problem described above we can derive that the expected number of stops is

$$\mathsf{E}Y = \sum_{i=1}^{n}\left[1 - \prod_{j=1}^{p}(1 - p_{ij})\right].$$

Table 8.3 Probabilities π_k and expectations $\mathsf{E}Y$.

k	$n=6,\ p=3$	$n=6,\ p=6$	$n=6,\ p=10$
1	0.028	0.000	0.000
2	0.417	0.020	0.000
3	0.555	0.231	0.019
4	0	0.502	0.203
5	0	0.231	0.506
6	0	0.016	0.272
$\mathsf{E}Y$	2.528	3.991	5.031

If we additionally assumed that the probabilities p_{ij} do not depend on j, so that $p_{ij} = p_i$, we would get

$$\mathsf{E}Y = \sum_{i=1}^{n} [1 - (1 - p_i)^p] \,.$$

Analyze the function

$$f(x) = 1 - (1 - x)^p, \qquad 0 < x < 1.$$

For $p \geq 2$, the function f is concave. If X is a random variable, then the *Jensen inequality* (see Rao 1973, Sect. 1e.5) says that for arbitrary convex function g the inequality $\mathsf{E}g(X) \geq g(\mathsf{E}X)$ holds. Consider the random variable X, which takes values $p_1, \ldots, p_n$, each with probability $1/n$. Since $p_1 + \cdots + p_n = 1$, we have $\mathsf{E}X = 1/n$. We already know that $\mathsf{E}f(X) \leq f(\mathsf{E}X)$. This leads to

$$\frac{1}{n} \sum_{i=1}^{n} [1 - (1 - p_i)^p] \leq \left[1 - \left(1 - \frac{1}{n}\right)^p\right].$$

We proved that in the case $p_{ij} = p_i$ the expected number of elevator stops is maximal when the probability of getting out is $1/n$ at each floor. On the other hand, it is clear that in the general case probabilities p_{ij} can be chosen in such a way that $\mathsf{E}Y$ can take arbitrary value between 1 and $\min(n, p)$.

8.4 MATCHING PAIRS OF SOCKS

Let n distinct *pairs of socks* be put into the laundry. Each of $2n$ socks has precisely one mate. When the laundry is returned, the socks are in a bag completely mixed. The customer draws out one sock at a time. Each sock is matched with its mate, if the mate has previously been drawn. Find a formula for the expected number of pairs formed after k socks have been drawn. This problem and several its solutions were published in *Am. Math. Monthly* (**95**, 1988, p. 188–189, Problem E3148).

It seems that the simplest solution is as follows. We label pairs of socks by the numbers 1 to n. The probability that the ith pair is drawn among the first k socks, is

$$\frac{\binom{2n-2}{k-2}}{\binom{2n}{k}}.$$

Introduce random variables $Y_1, \ldots, Y_n$ in the following way. Let

$$Y_i = \begin{cases} 1 & \text{if the } i\text{th pair is present among the first } k \text{ socks,} \\ 0 & \text{otherwise.} \end{cases}$$

The total number of pairs present among the first k socks is $\pi_k = \sum_{i=1}^{n} Y_i$, and its expectation is

$$\mathsf{E}\pi_k = \mathsf{E}\sum_{i=1}^{n} Y_i = \sum_{i=1}^{n} \mathsf{E}Y_i = n\frac{\binom{2n-2}{k-2}}{\binom{2n}{k}} = \frac{k(k-1)}{2(2n-1)}.$$

Assume now, that the customer places the socks successively on a table and then immediately removes each matching pair from the table. After k socks have been drawn, the expected number of socks lying on the table is

$$S_k = k - \mathsf{E}(2\pi_k) = \frac{2nk - k^2}{2n-1} = \frac{n^2 - (n-k)^2}{2n-1}.$$

It is obvious that S_k is maximal when $k = n$. We have in this case $S_n = n^2/(2n-1)$.

This problem was generalized to r-legged beings for $r > 2$. Assume that n distinct sets of socks be put into the laundry, where the ith set is from a creature with a_i legs, $i = 1, \ldots, n$. Then D. E. Knuth deduced that the expected number of complete sets of matching socks after k socks have been drawn at random is

$$\sum_{j=1}^{n} \frac{\binom{k}{a_j}}{\binom{a_1 + a_2 + \cdots + a_n}{a_j}}.$$

8.5 A GUESSING GAME

A broadcasting station prepared the following *guessing game* for its listeners. Every day a number from the set $\{1, 2, \ldots, n\}$ was randomly drawn. Listeners of this station know the number n. They can call to the station and guess the drawn number. A caller who guesses correctly receives a prize and the game is finished. If a caller guesses incorrectly, the station announces the number guessed and whether it is too high or too low. Find the expected number $f(n)$ of calls needed to correcty guess, assuming that each guess is made at random from the values that are not already excluded.

Let X be the drawn number. The probability that the first guess is correct is $1/n$. The probability that the first caller chooses the number k and that k is larger than X is equal to

$$\begin{aligned}&\mathsf{P}(\{\text{the caller chooses } k\} \cap \{k > X\})\\&\quad = \mathsf{P}(\text{the caller chooses } k) \times \mathsf{P}(k > X|\text{the caller chooses } k) = \frac{1}{n}\frac{k-1}{n}.\end{aligned}$$

The probability that the first caller chooses k and that k is smaller than X is $(n-k)/n^2$. This implies that

$$\begin{aligned}f(n) &= \frac{1}{n} + \sum_{k=2}^{n}[1 + f(k-1)]\frac{k-1}{n^2} + \sum_{k=1}^{n-1}[1 + f(n-k)]\frac{n-k}{n^2}\\&= 1 + \frac{2}{n^2}\sum_{k=1}^{n-1} k\, f(k).\end{aligned}$$

Therefore

$$n^2 f(n) = n^2 + 2\sum_{k=1}^{n-1} k\, f(k). \tag{8.1}$$

If we take $n-1$ instead of n, we get

$$(n-1)^2 f(n-1) = (n-1)^2 + 2\sum_{k=1}^{n-2} k\, f(k). \tag{8.2}$$

Subtracting (8.2) from (8.1), we obtain

$$n^2 f(n) - (n-1)^2 f(n-1) = n^2 - (n-1)^2 + 2(n-1) f(n-1),$$

which can be arranged into the form

$$f(n) = \frac{n^2-1}{n^2} f(n-1) + \frac{2n-1}{n^2}. \tag{8.3}$$

Define

$$g(n) = \frac{n}{n+1} f(n).$$

From (8.3) we find that $g(n)$ satisfies the equation

$$g(n) = g(n-1) + \frac{2n-1}{n(n+1)}. \tag{8.4}$$

Since $f(1) = 1$, we have an initial condition $g(1) = \frac{1}{2}f(1) = \frac{1}{2}$. Define $S_n = 1 + \frac{1}{2} + \cdots + \frac{1}{n}$. From (8.4) we derive

$$g(n) = \sum_{i=1}^{n} \frac{2i-1}{i(i+1)} = \sum_{i=1}^{n} \left(\frac{3}{i+1} - \frac{1}{i} \right) = 2S_n - \frac{3n}{n+1}.$$

This yields

$$f(n) = \frac{n+1}{n} g(n) = \frac{2(n+1)}{n} S_n - 3.$$

Using the Euler formula (4.4) for S_n, we finally obtain

$$f(n) = 2 \ln n + 2\gamma + o(1).$$

This problem was solved in *Am. Math. Monthly* (**100**, 1993, pp. 298–300, Problem E3448). A solution of the problem has been introduced in Wong (1964).

Of course, the expected number of calls could be smaller if the possible numbers are not chosen with the same probability. Such a model would be more realistic, but it would lead to more complicated results.

8.6 EXPECTED NUMBER OF DRAWS

Numbers are drawn randomly with replacement from the set $\{1, 2, \ldots, n\}$ until their sum first exceeds k such that $0 \le k \le n$. Find the expected *number of draws* E_k.

Assume that $n \ge 2$. It is clear that $E_0 = 1$. Let the outcome of the first draw be i. If $0 \le i \le k$, then we need the expected number of draws E_{k-i} so that the sum of drawn numbers first exceeds k. If $i > k$, no further draws are needed. This consideration leads to the equation

$$E_k = 1 + \frac{1}{n}(E_0 + E_1 + \cdots + E_{k-1}).$$

Using complete induction, it is easy to show that

$$E_k = \left(1 + \frac{1}{n}\right)^k, \qquad k = 0, 1, \ldots, n.$$

The case $k = n$ is considered most frequently (see *Am. Math. Monthly* **86**, 1979, p. 507). Then

$$E_n = \left(1 + \frac{1}{n}\right)^n,$$

which tends very rapidly to e as $n \to \infty$.

If the numbers are randomly selected from the set $\{0, 1, \ldots, n-1\}$ until their sum first exceeds k such that $0 \le k \le n-1$, the problem becomes more complicated. In this case we analogously obtain the formula

$$E_k = 1 + \frac{1}{n} \sum_{i=0}^{k} E_i. \tag{8.5}$$

In partikular, for $k = 0$ we have

$$E_0 = 1 + \frac{1}{n} E_0.$$

If we knew that E_0 is a finite number, we would immediately obtain

$$E_0 = \frac{n}{n-1}.$$

Since we have not proved it until now, we must use another procedure. The sum first exceeds zero in the mth draw ($m \geq 1$), if zero is selected $(m-1)$times, followed by an arbitrary positive number. This event has the probability

$$\left(\frac{1}{n}\right)^{m-1} \frac{n-1}{n}.$$

Thus

$$E_0 = \sum_{m=1}^{\infty} m \left(\frac{1}{n}\right)^{m-1} \frac{n-1}{n} = (n-1) \sum_{m=1}^{\infty} \frac{m}{n^m}.$$

Using formula (5.2) we really obtain $E_0 = n/(n-1)$. Since $E_k \leq (k+1)E_0$, we can see that also $E_1, \ldots, E_{n-1}$ are finite numbers. It follows from formula (8.5) that

$$E_k = \frac{n}{n-1} \left(1 + \frac{1}{n} \sum_{i=0}^{k-1} E_i\right)$$

and using complete induction, we obtain

$$E_k = \left(\frac{n}{n-1}\right)^{k+1}, \qquad k = 0, 1, \ldots, n-1.$$

Also, the case $k = n$ is most popular here.

Similar problem can be formulated for continuous rectangular distribution (see *Am. Math. Monthly* **68**, 1961, p. 18, Problem 3). Let $X_1, X_2, \ldots$ be a sample from the continuous rectangular distribution $\mathsf{R}(0, 1)$. Let N be a number such that $S_{N-1} \leq 1$, $S_N > 1$. Find $\mathsf{E}N$. However, this is the same problem as we solved in Section 7.4. It concerned fish, the weight of which was a random variable with distribution $\mathsf{R}(0, 1)$. The expected number of fish whose total weight exceeded 1 for the first time was e. Therefore, $\mathsf{E}N = e$.

8.7 LENGTH OF THE WIRE

Two points, say, X_1 and X_2, are randomly chosen in a segment of length $a > 0$. Find the expected *length of the wire* connecting points X_1 and X_2, specifically, $\mathsf{E}|X_1 - X_2|$.

We start with a more general problem. Let X_1 and X_2 be independent random variables with the same distribution function F and a density f. Define $X_{(1)} = \min(X_1, X_2)$, $X_{(2)} = \max(X_1, X_2)$. Calculate joint distribution function $G(x, y) = \mathsf{P}(X_{(1)} < x, X_{(2)} < y)$. Let $x < y$. Then

$$\begin{aligned}\{X_{(1)} < x, X_{(2)} < y)\} &= \{X_1 < x, X_2 < x\} \cup \{X_1 < x, x \leq X_2 < y\} \\ &\quad \cup \{X_2 < x, x \leq X_1 < y\}.\end{aligned}$$

Since the events on the right-hand side are disjoint and variables X_1, X_2 are independent, we have

$$G(x, y) = F^2(x) + 2F(x)[F(y) - F(x)], \qquad x < y.$$

If $x \geq y$, then

$$\{X_{(1)} < x, X_{(2)} < y)\} = \{X_1 < y, X_2 < y\},$$

so that $G(x, y) = F^2(y)$.

The joint density $g(x, y)$ of variables $X_{(1)}$ and $X_{(2)}$ is

$$g(x, y) = \frac{\partial^2 G(x, y)}{\partial x \partial y} = \begin{cases} 2f(x)f(y), & x < y, \\ 0, & x \geq y. \end{cases}$$

Hence

$$\mathsf{E}|X_1 - X_2| = \mathsf{E}(X_{(2)} - X_{(1)}) = \iint_{x<y} (y - x) 2f(x)f(y)\, \mathrm{d}x\, \mathrm{d}y.$$

If X_1 and X_2 have the rectangular distribution $\mathsf{R}(0, a)$, then $f(x) = 1/a$ for $0 < x < a$ and $f(x) = 0$ otherwise. Thus we obtain

$$\mathsf{E}|X_1 - X_2| = \int_0^a \left[\int_x^a (y - x) \frac{2}{a^2}\, \mathrm{d}y \right] \mathrm{d}x = \frac{a}{3}.$$

The expected wire length is $a/3$, which corresponds to an intuitively guessed result. Similarly, we get

$$\mathsf{E}(X_1 - X_2)^2 = \int_0^a \left[\int_x^a (y - x)^2 \frac{2}{a^2}\, \mathrm{d}y \right] \mathrm{d}x = \frac{a^2}{6}.$$

The last integral could be calculated from the previous results, since we have

$$\mathsf{E}(X_1 - X_2)^2 = 2\mathsf{E}X_1^2 - 2(\mathsf{E}X_1)^2.$$

The problem becomes more complicated if the wire joins two randomly and independently chosen points in a rectangle. Assume that the points have rectangular distribution in the rectangle.

The problem was solved independently by several authors. Its history and a tractable solution can be found in *SIAM Review* (**38**, 1996, p. 321–324). First, consider the rectangle $(0,1) \times (0,t)$. Let D be the length of the wire connecting two randomly chosen points in this rectangle. It was derived that

$$\begin{aligned} \mathsf{E}D &= \frac{\sinh^{-1} t}{6t} + \frac{t^2 \sinh^{-1} t^{-1}}{6} \\ &\quad - \frac{\sqrt{1+t^2}(t^4 - 3t^2 + 1)}{15t^2} + \frac{1}{15t^2} + \frac{t^3}{15}, \\ \mathsf{E}D^2 &= \frac{1}{6}(1+t^2). \end{aligned}$$

This formula can be simplified in the case that $t = 1$, when the rectangle is a square. Then

$$\begin{aligned} \mathsf{E}D &= \frac{1}{3}\ln(1+\sqrt{2}) + \frac{1}{15}(2+\sqrt{2}) = 0.5214, \\ \mathsf{E}D^2 &= \frac{1}{3}. \end{aligned}$$

If we have a rectangle with sides λ and $t\lambda$, and D_λ is the length of the wire connecting two randomly chosen points in the rectangle, then

$$\mathsf{E}D_\lambda = \lambda \mathsf{E}D, \qquad \mathsf{E}D_\lambda^2 = \lambda^2 \mathsf{E}D^2.$$

8.8 ANCIENT JEWISH GAME

The players use a dreidel instead of a dice. The dreidel is a four-sided top. Its sides are denoted by the Hebrew letters Nun, Gimel, Hay, and Shin. We use for simplicity N, G, H, and S. The game can be played with any number p of players. Each player contributes one coin to the pot before the game is started. The players successively spin the dreidel until some of them agree to stop. Assume that each letter has the same probability 1/4 to appear. The payoff to the spinning player is determined by outcomes in this way:

N : no payoff
G : entire pot
H : half the pot
S : puts a coin into the pot

When a player spins G and collects the entire pot, all players then contribute one coin to a new pot.

Feinerman (1976) states that the *game of dreidel* has been played by Jews for many years on the festival of Chanukah. We follow the analysis of the game of dreidel as presented in that paper.

The outcome G has a special role, since in some sense the game starts again following this outcome. For this reason we begin with the case that G does not appear.

Theorem 8.1 *Let Y_k be the amount in the pot at the start of the kth spin. Let g_{k-1} be the event that G did not occur in the first $k-1$ spins. Then*

$$\mathsf{E}(Y_k|g_{k-1}) = 2 + (p-2)\left(\frac{5}{6}\right)^{k-1}.$$

Proof. If $k = 1$, then the theorem holds, because p coins are in the pot before the first spin. Further, we use complete induction. Let the theorem hold for some $k \geq 1$. Since we assume that G does not appear in the kth spin, each of the remaining three letters has probability 1/3. Successively we obtain

$$\begin{aligned}\mathsf{E}(Y_{k+1}|g_k) &= \frac{1}{3}\mathsf{E}(Y_k|g_{k-1}) + \frac{1}{3}\cdot\frac{1}{2}\mathsf{E}(Y_k|g_{k-1}) + \frac{1}{3}[\mathsf{E}(Y_k|g_{k-1}) + 1] \\ &= \frac{5}{6}\mathsf{E}(Y_k|g_{k-1}) + \frac{1}{3} = \frac{5}{6}\left[2 + (p-2)\left(\frac{5}{6}\right)^{k-1}\right] + \frac{1}{3} \\ &= 2 + (p-2)\left(\frac{5}{6}\right)^{k}. \quad \square\end{aligned}$$

Now, we can formulate the main result.

Theorem 8.2 *Let X_n be the payoff on the nth spin. Then*

$$\mathsf{E}X_n = \frac{p}{4} + \frac{p-2}{8}\left(\frac{5}{8}\right)^{n-1}.$$

Proof. Assume first that G was at least once on some of the first $n-1$ spins. Let G_{n-k} be the event that the last G was on the $(n-k)$th spin, $1 \leq k \leq n-1$. This event has the probability

$$\pi_k = \frac{1}{4}\left(\frac{3}{4}\right)^{k-1}.$$

In this case we have

$$\begin{aligned}\mathsf{E}(X_n|G_{n-k}) &= \frac{1}{4}\cdot 0 + \frac{1}{4}\cdot\mathsf{E}(Y_k|g_{k-1}) + \frac{1}{4}\cdot\frac{1}{2}\mathsf{E}(Y_k|g_{k-1}) - \frac{1}{4}\cdot 1 \\ &= \frac{1}{4}\left[\frac{3}{2}\mathsf{E}(Y_k|g_{k-1}) - 1\right] = \frac{1}{4}\left[2 + \frac{3}{2}(p-2)\left(\frac{5}{6}\right)^{k-1}\right].\end{aligned}$$

Assume now that there were no previous Gs. This event has the probability

$$\pi_0 = \mathsf{P}(g_{n-1}) = \left(\frac{3}{4}\right)^{n-1}.$$

Then

$$\begin{aligned}\mathsf{E}(X_n|g_{n-1}) &= \frac{1}{4}\cdot 0 + \frac{1}{4}\cdot\mathsf{E}(Y_n|g_{n-1}) + \frac{1}{4}\cdot\frac{1}{2}\mathsf{E}(Y_n|g_{n-1}) - \frac{1}{4}\cdot 1 \\ &= \frac{1}{4}\left[\frac{3}{2}\mathsf{E}(Y_n|g_{n-1}) - 1\right] = \frac{1}{4}\left[2 + \frac{3}{2}(p-2)\left(\frac{5}{6}\right)^{n-1}\right].\end{aligned}$$

This implies

$$
\begin{aligned}
\mathsf{E}X_n &= \sum_{k=1}^{n-1} \mathsf{E}(X_n|G_{n-k})\pi_k + \mathsf{E}(X_n|g_{n-1})\pi_0 \\
&= \sum_{k=1}^{n-1} \frac{1}{16}\left(\frac{3}{4}\right)^{k-1}\left[2+\frac{3}{2}(p-2)\left(\frac{5}{6}\right)^{k-1}\right] \\
&\quad + \left(\frac{3}{4}\right)^{n-1}\frac{1}{4}\left[2+\frac{3}{2}(p-2)\left(\frac{5}{6}\right)^{n-1}\right] \\
&= \frac{p}{4}+\frac{p-2}{8}\left(\frac{5}{8}\right)^{n-1}. \quad \square
\end{aligned}
$$

The value $\mathsf{E}X_n$ does not depend on n only if $p = 2$. If $p \geq 3$, then $\mathsf{E}X_n$ is a decreasing function of n. We have

$$
\mathsf{E}X_1 = \frac{3p-2}{8}, \qquad \lim_{n\to\infty} \mathsf{E}X_n = \frac{p}{4}.
$$

Therefore, the first player has an advantage.

We should note that $\mathsf{E}X_n$ does not include the one-coin contribution, which each player puts in the pot after G is spined.

The game is not fair if we consider only the payoff. Maybe the players changed places from time to time. Maybe that they did not play only for a prize but mainly for amusement. Let us wish them it.

8.9 EXPECTED VALUE OF THE SMALLEST ELEMENT

Sophisticated tricks are used for calculating some expected values. The following problem was presented to participants of the International Mathematical Olympiad (see Lozansky and Rousseau 1996, p. 199).

Consider the set $\{1, 2, \ldots, n\}$. Let $1 \leq r \leq n$. There exist $N = \binom{n}{r}$ subsets with r elements. Choose randomly one from these subsets (each subset has the same probability to be chosen). Let ξ be the *smallest number* in this subset. Calculate $\mathsf{E}\xi$.

Define $M = \binom{n+1}{r+1}$. Let $X_1, \ldots, X_M$ be all subsets with $r+1$ elements from $\{0, 1, \ldots, n\}$ and let $Y_1, \ldots, Y_N$ be all subsets from $\{1, 2, \ldots, n\}$ with r elements.

Introduce matrix $\boldsymbol{A}_{M\times N} = (a_{ij})$ such that

$$
a_{ij} = \begin{cases} 1 & \text{if we get } Y_j \text{ after removing the smallest element from } X_i, \\ 0 & \text{otherwise.} \end{cases}
$$

The following two assertions are obvious:

1. On each row one element is 1 and all others are 0.

2. The number of 1 in the jth column is equal to the smallest element of Y_j.

Thus sum of the smallest elements of all sets $Y_1, \ldots, Y_N$ is the same as the number of rows of the matrix $\boldsymbol{A}$, namely, N. Then we have

$$\mathsf{E}\xi = \frac{M}{N} = \frac{n+1}{r+1}. \tag{8.6}$$

The same result can be derived even without the beautiful trick with the matrix $\boldsymbol{A}$, but the calculations are longer and more complicated. We apply the well-known formula

$$\sum_{\nu=0}^{s} \binom{\nu+k-1}{k-1} = \binom{s+k}{k}.$$

Let $1 \leq i \leq n-r+1$. The number of subsets with r elements of the set $\{1, 2, \ldots, n\}$, such that their smallest element is i, equals $\binom{n-i}{r-1}$. Hence sum of the smallest elements is

$$\begin{aligned} S &= \sum_{i=1}^{n-r+1} i\binom{n-i}{r-1} = \sum_{i=0}^{n-r}(n+1-r-i)\binom{i+r-1}{r-1} \\ &= (n+1)\sum_{i=0}^{n-r}\binom{i+r-1}{r-1} - \sum_{i=0}^{n-r}(i+r)\binom{i+r-1}{r-1} \\ &= (n+1)\binom{n}{r} - r\sum_{i=0}^{n-r}\binom{i+r}{r} = (n+1)\binom{n}{r} - r\binom{n+1}{r+1} \\ &= \binom{n+1}{r+1} = M. \end{aligned}$$

We know that the number of subsets with r elements of the set $\{1, 2, \ldots, n\}$ is $N = \binom{n}{r}$. This also implies formula (8.6).

As an illustration of our result, we introduce a solution of the following problem. A pack of the usual 32 playing cards is thoroughly shuffled and then turned so that the cards are face down. Then one card after the other is successively turned until an ace is found. Find the expected number of cards that must be turned.

If the cards are well shuffled, then each subset with four elements of the set $\{1, 2, \ldots, 32\}$ has the same probability of containing orders of aces. The ace having the smallest order will be found first. We have $n = 32$, $r = 4$, and thus the expected number of turned cards is

$$\frac{n+1}{r+1} = \frac{33}{5} = 6.6.$$

8.10 BALLOT COUNT

An election involved only two contestants, A and B. We know in advance that each of them received N of $2N$ votes cast. This means that the election will be deadlocked after all the votes are counted. The *ballots* during count are randomly selected from a single ballot box. When k ballots are drawn, A has A_k and B has B_k votes. Of course, $A_k + B_k = k$. Define $X_k = |A_k - B_k|$. Find $\mathsf{E}X_k$. (See Problem 10248, "early returns in a tied election," *Am. Math. Monthly* **102**, 1995, pp. 554–556.)

Let $\lfloor x \rfloor$ be the integer part of the number x. It is clear that X_k takes one of the values $k, k-2, k-4, \ldots$. The variable X_k equals $k - 2i$ ($i = 0, 1, \ldots, \lfloor k/2 \rfloor$) either if $A_k = i$, $B_k = k - i$ or if $A_k = k - i$, $B_k = i$. Obviously, we have for arbitrary nonnegative integers i, j satisfying $i + j = k$ that

$$\mathsf{P}(A_k = i, B_k = j) = \frac{\binom{N}{i}\binom{N}{j}}{\binom{2N}{i+j}}.$$

If $i \neq k - i$, then

$$\mathsf{P}(X_k = k - 2i) = 2\frac{\binom{N}{i}\binom{N}{k-i}}{\binom{2N}{k}}.$$

If $i = k - i$, we get $i = k/2$. This case occurs only if k is even. However, for $i = k/2$, we obtain $X_k = 0$ and

$$\mathsf{P}(X_k = 0) = \frac{\binom{N}{k/2}^2}{2Nk}.$$

Calculate

$$\mathsf{E}X_k = \sum_{i=0}^{\lfloor k/2 \rfloor} (k - 2i)\mathsf{P}(X_k = k - 2i).$$

If k is even and $i = k/2$, then it is immaterial whether the number $k - 2i$ is multiplied by $\mathsf{P}(X_k = 0)$ or by any other expression, since the result is zero in all cases. Thus

$$\mathsf{E}X_k = \frac{2\sum_{i=0}^{\lfloor k/2 \rfloor}(k-2i)\binom{N}{i}\binom{N}{k-i}}{\binom{2N}{k}}.$$

Further calculations are based on the formula

$$(j-i)\binom{N}{i}\binom{N}{j} = N\left[\binom{N-1}{i}\binom{N-1}{j-1} - \binom{N-1}{i-1}\binom{N-1}{j}\right].$$

This formula can be easily verified, since

$$\begin{aligned} j\binom{N}{i}\binom{N}{j} &= N\left[\binom{N-1}{i}\binom{N-1}{j-1} - \binom{N-1}{i-1}\binom{N-1}{j-1}\right], \\ i\binom{N}{i}\binom{N}{j} &= N\left[\binom{N-1}{i-1}\binom{N-1}{j} - \binom{N-1}{i-1}\binom{N-1}{j-1}\right]. \end{aligned}$$

Here we define $\binom{m}{-1} = 0$. Hence

$$\mathsf{E}X_k = \frac{2N \sum_{i=0}^{\lfloor k/2 \rfloor} \left[\binom{N-1}{i}\binom{N-1}{k-i-1} - \binom{N-1}{i-1}\binom{N-1}{k-i} \right]}{\binom{2N}{k}} = \frac{2N \binom{N-1}{\lfloor k/2 \rfloor}\binom{N-1}{\lfloor k/2 \rfloor - 1}}{\binom{2N}{k}}.$$

Finally we obtain

$$\mathsf{E}X_k = \begin{cases} \dfrac{k(2N-k)}{2N} \dbinom{N}{k/2}^2 \Big/ \dbinom{2N}{k} & \text{for } k \text{ even,} \\ \dfrac{k(2N-k+1)}{2N} \dbinom{N}{(k-1)/2}^2 \Big/ \dbinom{2N}{k-1} & \text{for } k \text{ odd.} \end{cases}$$

Let n be a positive integer. The case $N = k = 2n$ is most popular. Then

$$\mathsf{E}X_{2n} = \frac{n \binom{2n}{n}^2}{\binom{4n}{2n}}.$$

If n is a large number, we can apply the *Stirling formula*

$$n! = \left(\frac{n}{e}\right)^n \sqrt{2\pi n} \left(1 + \frac{1}{12n} + \frac{1}{288n^2} + \cdots \right).$$

The result is $\mathsf{E}X_{2n} \doteq \sqrt{2n/\pi}$. For $N = 100$, we get $\mathsf{E}X_{100} \doteq 6$; for $N = 5,000,000$ we obtain $\mathsf{E}X_{5,000,000} \doteq 1262$.

Now, further consider problems connected with the problem formulated above. Find the probability that during ballot random selections, the number of votes for A is never smaller than those for B. Our solution can be based on results derived in Section 3.2 using the *reflection method*. Selected ballots can be represented as a queue of length $2N$, where N customers are labeled A and N customers are labeled B. If B never has more votes than A, this corresponds to the case where a ticket-vending automat having $a = 0$ ballots in its store at the beginning of operation will not become blocked. Using formula (3.1), we find that the probability of this event is

$$P_0(2N, N) = 1 - \frac{N}{N+1} = \frac{1}{N+1}.$$

We have the following remark to this result. We know that the number of all sequences of ballots is $\binom{2N}{N}$. The number of sequences in which B sometimes has more votes than A does equal the number of routes from point A to point C in Fig. 3.2, specifically, $\binom{2N}{N+1}$. Then number of the sequences when B never has more votes than A is

$$\binom{2N}{N} - \binom{2N}{N+1} = \frac{1}{N+1}\binom{2N}{N} = C_N.$$

Here C_N is the Nth *Catalan number*.

Lozansky and Rousseau (1996, p. 186 and p. 230) introduce further problems concerning the ballot count. We introduce one of them. Find the number of sequences of ballots such that A is the leader of all the time except for the last vote.

Each such sequence must start with a vote for A and finish with a vote for B. Removing the first and the last votes, we obtain a sequence of the length $2N - 2$ with $N - 1$ votes for A and $N - 1$ votes for B. In this subsequence B must never be the leader. The number of such subsequences is known from the preceding analysis and equals C_{N-1}. Therefore, there exist C_{N-1} sequences of ballots such that A leaders except for the last vote.

8.11 BERNOULLI PROBLEM

In a town m married couples of the same age were selected. After a few years only a survivors were found from those $2m$ persons. Assume that each individual is alive with the same probability independently of the others. Find the expected number of surviving married couples. This *Bernoulli problem* was solved by Daniel Bernoulli in 1768.

Assume that $a \geq 2$, otherwise the problem is trivial. Let I_j be the *indicator* of surviving of the jth married couple. This means that $I_j = 1$ if both partners of the jth couple are alive and $I_j = 0$ otherwise. Let S_a be number of surviving couples. Then

$$\mathsf{E}S_a = \mathsf{E}\sum_{j=1}^{m} I_j = m\mathsf{E}I_1 = m\mathsf{P}(I_1 = 1).$$

The a surviving individuals from $2m$ people can be chosen in $\binom{2m}{a}$ ways. The number of ways, when both partners from the first couple are alive, is $\binom{2m-2}{a-2}$. Thus we have

$$\mathsf{P}(I_1 = 1) = \frac{\binom{2m-2}{a-2}}{\binom{2m}{a}} = \frac{a(a-1)}{2m(2m-1)}.$$

We proved that

$$\mathsf{E}S_a = \frac{a(a-1)}{2(2m-1)}.$$

If we started with 100 married couples and only 100 individuals are alive after some time, then the expected number of surviving couples in which both partners are alive would be $\mathsf{E}S_{100} = 24.87$.

This procedure and other two methods how to derive the formula for $\mathsf{E}S_a$, are introduced in Stirzaker (1994, pp. 158–159).

A *problem concerning a committee* (see Gordon 1997, p. 101, Example 1) can be solved similarly. Six persons are randomly chosen from six married couples in a committee for degustation of wine. Find the expected number of couples who participate in the committee. Define

$$X_i = \begin{cases} 1 & \text{if the } i\text{th couple is in the committee,} \\ 0 & \text{otherwise.} \end{cases}$$

Then $X = X_1 + \cdots + X_6$ is the number of couples in the committee. Since

$$\mathsf{P}(X_i = 0) = \frac{\binom{10}{6}}{\binom{12}{6}} = \frac{5}{22},$$

we have

$$\mathsf{E}X_i = \mathsf{P}(X_i = 1) = 1 - \mathsf{P}(X_i = 0) = \frac{17}{22}.$$

Thus the result is

$$\mathsf{E}X = \sum_{i=1}^{6} \mathsf{E}X_i = 6 \times \frac{17}{22} = 4.64.$$

8.12 EQUAL NUMBERS OF HEADS AND TAILS

A fair coin is tossed $(2n)$times. The number of heads and tails is recorded. Let T be a random variable that is equal to the number of cases such that the number of heads and tails are equal. The initial state (no head and no tail) is not taken into account. Find the expectation of T.

It is clear that the same *number of heads and tails* can occur only in an even number of tosses. We introduce *indicators*

$$T_{2k} = \begin{cases} 1 & \text{if number of heads and tails is the same after } 2k \text{ tosses,} \\ 0 & \text{otherwise,} \end{cases}$$

where $k = 1, 2, \ldots, n$ (see solution of Problem 10, 355 in *Am. Math. Monthly* **104**, 1997, pp. 175–176). Further, we can see that

$$u_{2k} = \mathsf{P}(T_{2k} = 1) = \binom{2k}{k}\left(\frac{1}{2}\right)^k \left(\frac{1}{2}\right)^k = \binom{2k}{k}\frac{1}{4^k}.$$

Since $T = T_2 + T_4 + \cdots + T_{2n}$ a $\mathsf{E}T_{2k} = u_{2k}$, we have

$$\mathsf{E}T = \sum_{k=1}^{n} \mathsf{E}T_{2k} = \sum_{k=1}^{n} \binom{2k}{k}\frac{1}{4^k} = (n+1)\binom{2n+1}{n}\frac{1}{4^n} - 1. \qquad (8.7)$$

Table 8.4 Values ET and their approximations.

n	ET	Approximation of ET	n	ET	Approximation of ET
10	2.700	2.656	100	10.326	10.312
50	7.039	7.019	500	24.250	24.244

The last assertion can be proved by complete induction. If n is large, we can use the Stirling formula (8.7) and obtain an approximation

$$\mathsf{E}T \doteq \sqrt{\frac{4n+2}{\pi}} - 1.$$

The quality of this approximation can be evaluated in Tab. 8.4.

We introduce a few remarks on this topic. We know that u_{2k} is the probability that after $2k$ tosses the number of heads and tails will be the same. The probability that this number will be the same after $2k$ tosses for the first time, is

$$f_{2k} = \frac{1}{2k} u_{2k-2}, \qquad k = 1, 2, \ldots, n$$

(see Feller 1964, Vol. I, Chap. III, Sect. 4). The probability that in the $2n$ tosses number of heads and tails will be the same $r$times is

$$z_{2n}^{(r)} = \frac{1}{2^{2n-r}} \binom{2n-r}{n}, \qquad n \geq 1$$

(see Feller 1964, Vol. I, Chap. III, Sect. 6). This leads to another formula:

$$\mathsf{E}T = \sum_{r=0}^{n} \frac{r}{2^{2n-r}} \binom{2n-r}{n}.$$

Riordan (1968, p. 31), shows how this formula can be rearranged into form (8.7). In the same publication the variance of the variable T

$$\mathsf{var}\, T = 2n - 3\mathsf{E}T + (\mathsf{E}T)^2$$

is derived. Higher moments can be obtained from explicit formulas for factorial moments published in Kirschenhofer and Prodinger (1994).

Further, using the Stirling formula it can be proved that $\sqrt{\pi k}\, u_{2k} \to 1$ as $k \to \infty$. For large values of k, we can use the approximation

$$u_{2k} \doteq \frac{1}{\sqrt{\pi k}}.$$

8.13 PEARLS

There are a white and b black *pearls* in a jewel box. It is known that $a > 0$, $b > 0$. A pearl is chosen randomly. If it is black, the process terminates. If it is white, the pearl is returned and another white pearl is added to the box. Find the probability that the process terminates in the kth step. Calculate the probability that the process terminates in finitely many steps. Find the expected number of drawn pearls.

We start with an auxiliary assertion. Let a random variable X take values $1, 2, \ldots$ with probabilities $\mathsf{P}(X = k) = p_k$. Since p_k are nonnegative numbers, we can write

$$\begin{aligned} \mathsf{E}X &= \sum_{k=1}^{\infty} k p_k \\ &= p_1 \\ &\quad + p_2 + p_2 \\ &\quad + p_3 + p_3 + p_3 \\ &\quad + \cdots \\ &= (p_1 + p_2 + p_3 + \cdots) + (p_2 + p_3 + \cdots) + (p_3 + \cdots) + \cdots \\ &= \sum_{k=0}^{\infty} \mathsf{P}(X > k). \end{aligned} \tag{8.8}$$

In this connection another auxiliary assertion is useful. Assume that $\mathsf{E}X < \infty$ also holds for the abovementioned random variable X. This means that $\sum_{i=1}^{\infty} i p_i < \infty$. Then we can see that

$$0 \le k\mathsf{P}(X > k) = k \sum_{i=k+1}^{\infty} p_i \le \sum_{i=k+1}^{\infty} i p_i$$

holds for every nonnegative integer k. The right-hand side tends to zero as $k \to \infty$. Thus we proved that the assumption implies

$$k\mathsf{P}(X > k) \to 0 \qquad \text{as} \quad k \to \infty. \tag{8.9}$$

In our case X is the random variable that is equal to number of drawn pearls (i.e., the number of trials). It is clear that $\mathsf{P}(X > 0) = 1$. The sampling process does not finish after the first step with probability

$$\mathsf{P}(X > 1) = \frac{a}{a + b}.$$

The conditional probability that the sampling does not finish in the ith step, given that it has not finished before, is equal to

$$\frac{a + i - 1}{a + b + i - 1}.$$

Thus the probability, that sampling does not finish in the first k steps ($k = 1, 2, \ldots$) is

$$\mathsf{P}(X > k) = \frac{a}{a + b} \frac{a + 1}{a + b + 1} \cdots \frac{a + k - 1}{a + b + k - 1} = \frac{(a + b - 1)!\,(a + k - 1)!}{(a - 1)!\,(a + b + k - 1)!}.$$

It is easy to verify that $\mathsf{P}(X > k) \to 0$ as $k \to \infty$. Thus sampling finishes in finitely many steps with probability 1.

The probability that sampling finishes in the kth step is

$$\begin{aligned}\mathsf{P}(X = k) &= \mathsf{P}(X > k-1) - \mathsf{P}(X > k) \\ &= \frac{a}{a+b}\frac{a+1}{a+b+1}\cdots\frac{a+k-2}{a+b+k-2}\frac{b}{a+b+k-1}.\end{aligned}$$

If $b = 1$, then

$$k\mathsf{P}(X > k) = k\frac{a!\,(a+k-1)!}{(a-1)!\,(a+k)!} = \frac{ak}{a+k},$$

which tends to $a > 0$ as $k \to \infty$. Since (8.9) does not hold, it follows that $\mathsf{E}X = \infty$.

Let $b \geq 2$. Then we find from (8.8) that

$$\begin{aligned}\mathsf{E}X &= \sum_{k=0}^{\infty} \mathsf{P}(X > k) \\ &= 1 + \frac{(a+b-1)!}{(a-1)!\,(b-1)!}\sum_{k=1}^{\infty}\left[\frac{(a+k-1)!}{(a+b+k-2)!} - \frac{(a+k)!}{(a+b+k-1)!}\right].\end{aligned}$$

The sum has a telescopic character; its terms cancel successively. Thus

$$\mathsf{E}X = 1 + \frac{(a+b-1)!}{(a-1)!\,(b-1)}\frac{a!}{(a+b-1)!} = 1 + \frac{a}{b-1} = \frac{a+b-1}{b-1}.$$

The results are rather surprizing. If we have one white and one black pearl at the beginning in the jewel box, the expected number of draws is infinite. If the jewel box contains a million white pearls and only two black pearls, the expected number of draws is finite! At the same time, in both cases sampling stops in a finite number of steps with probability 1.

Assume that $a > 0$, $b = 1$ and that the sampling procedure was realized m times independently. Let $X_1, \ldots, X_m$ be numbers of draws until a black pearl is sampled. Define $\bar{X} = (X_1 + \cdots + X_m)/m$. It can be proved (see *Am. Math. Monthly* **105**, 1998, pp. 181–182, Problem 10504 called "another Pólya urn scheme") that we have for every $\varepsilon > 0$

$$\mathsf{P}\left\{\left|\frac{\bar{X}}{\ln m} - a\right| > \varepsilon\right\} \to 0 \qquad \text{for} \qquad m \to \infty.$$

This means that in the case $b = 1$ the variable $\bar{X}/\ln m$ can be used as an estimate for parameter a.

9

Problems on statistical methods

9.1 PROOFREADING

A manuscript with an unknown number of misprints is subjected to a *proofreading* in an effort to detect and correct as much misprints as possible — in an ideal case all the misprints. The proofreading is usually done by the author or by the translator. In some cases the author acts as the proofreader alone, in other cases with the assistance of the author's colleagues or students. If the manuscript is extensive, it is nearly impossible to detect all misprints (and we do not even mention subject errors). A few statistical models of proofreading were proposed in scientific papers.

Model I. Two proofreaders, A and B, say, read independently of each other a manuscript, which contains an unknown number M of misprints. Assume that a misprints were noticed by A and b misprints by B. Comparing their proofsheets, they found that c misprints were noticed by both. Let p and q be the probability that proofreader A and B notice any given misprint, respectively. Then the number of misprints noticed by A and B has the binomial distribution $\mathrm{Bi}(M, p)$ and $\mathrm{Bi}(M, q)$, respectively. The number of misprints noticed by both has distribution $\mathrm{Bi}(M, pq)$. Parameters M, p, and q are unknown. They can be estimated using the *moment method* (see Pólya 1976, Liu 1988). We construct a system of equation such that the observed values are equal to their expectations. The expectation of a random variable with binomial distribution $\mathrm{Bi}(n, \pi)$ is $n\pi$. Thus in our case we obtain the system

$$Mp = a, \qquad Mq = b, \qquad Mpq = c.$$

Its solution is denoted by the same symbols complemented by a hat, a star, an so on because, in fact, we do not obtain the true values of the parameters but only their estimates. Here we have

$$M = \frac{Mp \times Mq}{Mpq} = \frac{ab}{c}, \qquad p = \frac{a}{M} = \frac{c}{b}, \qquad q = \frac{b}{M} = \frac{c}{a}$$

and so we can write

$$\hat{M} = \frac{ab}{c}, \qquad \hat{p} = \frac{c}{b}, \qquad \hat{q} = \frac{c}{a}.$$

An important characteristic is the number of unnoticed misprints M_1, which were overlooked by both proofreaders. It is clear that

$$M_1 = M - a - b + c.$$

We obtain an estimate $\hat{M}_1$ for M_1 by inserting $\hat{M}$ for M. This yields

$$\hat{M}_1 = \frac{ab}{c} - a - b + c = \frac{(a-c)(b-c)}{c}.$$

My colleagues I. Netuka and J. Veselý translated the book by Rudin (1974) into Czech. They read proofsheets independently and using the method described above estimated that approximately eight misprints were unnoticed and uncorrected. For more than 20 years the book has been intensively used as a textbook by many students and teachers, who found seven misprints. Of course, some other misprints can remain still unnoticed; nevertheless the data show that the model can be useful.

Model II. A proofreader reads a manuscript twice independently: once from the first proofsheet and later after a long pause from the second proofsheet. Let X_1 and X_2 denote the number of misprints observed in the first and second proofreadings, respectively. Let X_{12} be the number of misprints noticed in both proofreadings. Assume that M is the total number of misprints in the manuscript (noticed or unnoticed) and that p is the probability that the proofreader notices any given misprint. In fact, we have a special case of the model I with $p = q$. Then the parameter M can be estimated by

$$M^* = \frac{X_1 X_2}{X_{12}},$$

and the parameter p can be estimated by the average of estimates for p and q derived in model I, namely, by

$$p^* = \frac{1}{2}\left(\frac{X_{12}}{X_2} + \frac{X_{12}}{X_1}\right) = \frac{X_{12}(X_1 + X_2)}{2X_1 X_2}.$$

These formulas are introduced in Liu (1988). A statistician would prefer other estimates. Since $X_1 \sim \mathrm{Bi}(M,p)$, $X_2 \sim \mathrm{Bi}(M,p)$ and independence of X_1 and X_2 is assumed, it is not difficult to prove that $X_1 + X_2 \sim \mathrm{Bi}(2M,p)$. Because $X_{12} \sim \mathrm{Bi}(M,p^2)$, the moment method yields the following system of equations:

$$2Mp = X_1 + X_2, \qquad Mp^2 = X_{12}.$$

Their solution is

$$\hat{M} = \frac{(X_1 + X_2)^2}{4X_{12}}, \qquad \hat{p} = \frac{2X_{12}}{X_1 + X_2}.$$

Model III. A proofreader reads the same proofsheet twice. This means that before the second proofreading all the misprints found before are already marked and corrected. Assume again that M is the total number of misprints in the manuscript (before the first reading) and that p is the probability that the proofreader notices any given misprint. Let X_1 and X_2 denote the number of misprints observed in the first and second proofreadings, respectively. We have $X_1 \sim \mathsf{Bi}(M, p)$, $X_2 \sim \mathsf{Bi}(M - X_1, p)$, and the moment method yields equations

$$Mp = X_1, \qquad (M - X_1)p = X_2.$$

The solution is

$$\hat{M} = \frac{X_1^2}{X_1 - X_2}, \qquad \hat{p} = \frac{X_1 - X_2}{X_1}.$$

Some papers (e.g., Ferguson and Hardwick 1989) are devoted to the case when a manuscript is subjected to a series of proofreadings. Each proofreading costs an amount and also each overlooked misprint costs a certain amount. After the first proofreading, many misprints can be overlooked. On the other hand, a large number of proofreadings can be more expensive than the price paid for the remaining overlooked misprints. Optimal stopping rules were derived. These problems are important in connection with debugging and testing computer software.

9.2 HOW TO ENHANCE THE ACCURACY OF A MEASUREMENT

It is known that our instruments do not measure absolutely accurately. Assume that we have an instrument to measure lengths. If an object has length c, the result of measurement is a random variable X such that

$$\mathsf{E}X = c, \qquad \mathsf{var}\, X = \sigma^2.$$

Assume that variance σ^2 does not depend on the true value c. We want to measure the lengths of two sticks. One of them is obviously longer than the other. We are allowed to measure only twice.

The simplest procedure is to measure the length a of the longer stick and then the length b of the shorter stick. Our assumptions guarantee that the result of each measurement is a random variable with variance σ^2. The following method gives a more precise result.

If we put two sticks together, we can measure the sum of their lengths $s = a + b$ as well as their difference $d = a - b$. If we knew s and d exactly, we would get

$$a = \frac{s + d}{2}, \qquad b = \frac{s - d}{2}.$$

Let S and D be our measurement of the sum and difference, respectively. Assume that the measurements are independent. Then we have

$$\mathsf{E}S = s, \qquad \mathsf{E}D = d, \qquad \operatorname{var} S = \sigma^2, \qquad \operatorname{var} D = \sigma^2.$$

It is natural to take $A = (S + D)/2$ as an estimate of the length of the longer stick and $B = (S - D)/2$ as an estimate of the length of the shorter stick. It is easy to see that

$$\mathsf{E}A = a, \qquad \mathsf{E}B = b, \qquad \operatorname{var} A = \operatorname{var} B = \frac{\operatorname{var} S + \operatorname{var} D}{4} = \frac{\sigma^2}{2}.$$

These results are more accurate since the variance of each estimate is only $\sigma^2/2$ instead of σ^2, which we obtained using the first method.

Technically we must be sure that no additional inaccuracy arises when the sticks are put together.

This problem is frequently presented in introductory courses of mathematical statistics. It can be found also in the book by Mosteller (1965, Problem 49).

9.3 HOW TO DETERMINE THE AREA OF A SQUARE

A physicist measured twice a value (distance, velocity, etc.). We denote the true value by μ. The results of measurement were $X_1 = 3.1$ and $X_2 = 3.3$, say. However, a formula contains, in fact, μ^2. It seems that two methods could be used for statistical estimation of μ^2.

First method: Calculate the average $\bar{X} = (X_1 + X_2)/2 = 3.2$ and estimate μ^2 by $3.2^2 = 10.24$.

Second method: Calculate the square of each measurement and estimate μ^2 by the average of these two squares. Since $X_1^2 = 9.61$ and $X_2^2 = 10.89$, this estimate of μ^2 is $(9.61 + 10.89)/2 = 10.25$.

Which of the methods is better? Which procedure would a professional statistician use? Before solving the problems, we introduce an auxiliary assertion.

Theorem 9.1 *Let* $Y \sim \mathsf{N}(\nu, \delta^2)$. *Then* $\operatorname{var} Y^2 = 2\delta^4 + 4\delta^2\nu^2$.

Proof. We have

$$\begin{aligned} \mathsf{E}Y^2 &= \operatorname{var} Y + (\mathsf{E}Y)^2 = \delta^2 + \nu^2, \\ \mathsf{E}Y^4 &= \mathsf{E}[(Y - \nu) + \nu]^4 = 3\delta^4 + 6\nu^2\delta^2 + \nu^4, \\ \operatorname{var} Y^2 &= \mathsf{E}Y^4 - (\mathsf{E}Y^2)^2 = 2\delta^4 + 4\delta^2\nu^2. \quad \square \end{aligned}$$

Consider a general version of the problem. The physicist has n independent measurements $X_1, \ldots, X_n$ of the value μ. Let $X_1, \ldots, X_n$ be a sample from $\mathsf{N}(\mu, \sigma^2)$, where the parameters μ and σ^2 are not known. Define

$$\bar{X} = \frac{1}{n}\sum_{i=1}^{n} X_i, \qquad S^2 = \frac{1}{n-1}\sum_{i=1}^{n}(X_i - \bar{X})^2.$$

The first method uses $\xi = \bar{X}^2$ as an estimate for μ^2. Since $\bar{X} \sim \mathsf{N}(\mu, \sigma^2/n)$, we get

$$\mathsf{E}\xi = \mathsf{E}\bar{X}^2 = \mathsf{var}\,\bar{X} + (\mathsf{E}\bar{X})^2 = \frac{\sigma^2}{n} + \mu^2,$$

and in view of Theorem 9.1 we obtain

$$\mathsf{var}\,\xi = \mathsf{var}\,\bar{X}^2 = \frac{2\sigma^4}{n^2} + \frac{4\sigma^2\mu^2}{n}. \tag{9.1}$$

We can see that $\bar{X}^2$ is not an unbiased estimate, since $\mathsf{E}\xi$ is larger than the wanted value μ^2 by an amount σ^2/n . The accuracy of an estimate is measured by the *mean square error* (MSE), which is

$$\mathsf{MSE}_\xi = \mathsf{E}(\xi - \mu^2)^2 = \mathsf{var}\,\xi + (\mathsf{E}\xi - \mu^2)^2 = \frac{3\sigma^4}{n^2} + \frac{4\sigma^2\mu^2}{n}.$$

In the second method μ^2 is estimated by $\eta = (1/n)\sum_{i=1}^n X_i^2$. Using Theorem 9.1, we obtain

$$\mathsf{E}\eta = \frac{1}{n}\sum_{i=1}^n \mathsf{E}X_i^2 = \sigma^2 + \mu^2, \qquad \mathsf{var}\,\eta = \frac{1}{n^2}\sum_{i=1}^n \mathsf{var}\,X_i^2 = \frac{2\sigma^4 + 4\sigma^2\mu^2}{n}.$$

The bias of the estimate η is σ^2. If $n \geq 2$, then it is larger than the bias of the estimate ξ, which is σ^2/n. The mean square error is

$$\mathsf{MSE}_\eta = \mathsf{E}(\eta - \mu^2)^2 = \mathsf{var}\,\eta + (\mathsf{E}\eta - \mu^2)^2 = \frac{2\sigma^4}{n} + \frac{4\sigma^2\mu^2}{n} + \sigma^4.$$

After some computation we get

$$\mathsf{MSE}_\eta - \mathsf{MSE}_\xi = \frac{(n+3)(n-1)}{n^2}\sigma^4.$$

If $n = 1$, then both estimates ξ and η are identical and equally accurate. The estimate ξ is more accurate than η when $n \geq 2$.

A statistician would know that $\bar{X}$ and S^2 is a *complete minimal sufficient statistic* for μ and σ^2 (see Kendall and Stuart, 1973, Vol. II, Chap. 23). The statistician would estimate μ^2 by a function of $\bar{X}$ and S^2 such that its expectation would be μ^2. It is easy to find that in our case the estimate

$$\zeta = \bar{X}^2 - \frac{1}{n}S^2$$

has the above mentioned properties. The estimate ζ is unbiased, and thus its mean square error is

$$\mathsf{MSE}_\zeta = \mathsf{E}\left(\bar{X}^2 - \frac{1}{n}S^2 - \mu^2\right)^2 = \mathsf{var}\left(\bar{X}^2 - \frac{1}{n}S^2\right).$$

Since we have a sample from a normal distribution, variables $\bar{X}$ and S^2 are independent. Thus $\text{var}(\bar{X}^2+n^{-1}S^2) = \text{var}\,\bar{X}^2+n^{-2}\text{var}\,S^2$. The value $\text{var}\,\bar{X}^2$ is introduced in formula (9.1). It is known that $(n-1)S^2/\sigma^2 \sim \chi^2_{n-1}$ and that the variance of a variable with χ^2 distribution is 2times the number of degrees of freedom. Thus

$$\frac{1}{n^2}\,\text{var}\,S^2 = \frac{\sigma^4}{n^2(n-1)^2}\,\text{var}\frac{(n-1)S^2}{\sigma^2} = \frac{\sigma^4}{n^2(n-1)^2}2(n-1) = \frac{2\sigma^4}{n^2(n-1)}.$$

This gives

$$\text{MSE}_\zeta = \text{var}\,\zeta = \frac{2\sigma^4}{n(n-1)} + \frac{4\mu^2\sigma^2}{n}.$$

To compare the accuracies of the estimates ζ and ξ, we calculate

$$\text{MSE}_\xi - \text{MSE}_\zeta = \frac{n-3}{n^2(n-1)}\sigma^4.$$

When $n > 3$ the estimate ζ is more accurate than ξ. Both estimates are equally accurate when $n = 3$ and ξ is more accurate than ζ when $n = 2$ (however, the estimate ξ is biased, whereas ζ is unbiased).

We started with a situation in which only two values, X_1 and X_2 were available. A crazy idea is to use the product X_1X_2 as an estimate of μ^2. We investigate properties of this estimate. It is unbiased, since the independence ensures $\text{E}X_1X_2 = \text{E}X_1\text{E}X_2 = \mu^2$. The mean square error of the estimate is the same as its variance:

$$\begin{aligned} \text{var}\,X_1X_2 &= \text{E}X_1^2X_2^2 - (\text{E}X_1)^2(\text{E}X_2)^2 = (\text{E}X_1^2)(\text{E}X_2^2) - \mu^4 \\ &= (\sigma^2+\mu^2)^2 - \mu^4 = \sigma^4 + 2\mu^2\sigma^2. \end{aligned}$$

Surprisingly, we found that $\text{var}\,X_1X_2 = \text{MSE}_\zeta$ (since $n = 2$). However, this equality clearly holds, because for $n = 2$ we have

$$\begin{aligned} \zeta &= \bar{X}^2 - \frac{1}{n}S^2 \\ &= \left(\frac{X_1+X_2}{2}\right)^2 - \frac{1}{2}\left[\left(X_1 - \frac{X_1+X_2}{2}\right)^2 + \left(X_2 - \frac{X_1+X_2}{2}\right)^2\right] \\ &= \frac{1}{4}\left(\frac{X_1+X_2}{2}\right)^2 - \frac{1}{4}\left(\frac{X_1-X_2}{2}\right)^2 = X_1X_2, \end{aligned}$$

thus, both estimates ζ and X_1X_2 are identical.

One can guess that something similar may also hold for general n. In fact, it is easy to verify that

$$\zeta = \bar{X}^2 - \frac{1}{n}S^2 = \frac{1}{n(n-1)}\sum_{1\le i\ne j\le n}\sum X_iX_j.$$

Thus the estimate ζ is the average of all possible products X_iX_j with different indices $i \ne j$.

Table 9.1 Travel times (in minutes) needed to reach the airport.

Route 1	$34\frac{1}{2}$	35	34	$34\frac{1}{2}$
Route 2	33	32	19	34

9.4 TWO ROUTES TO THE AIRPORT

A guest from abroad visited a university. After the visit, a colleague from the university is going to take the guest by a car to the *airport*. There are two routes to the airport. Each of them was used 4 times, and travel times are recorded in Table 9.1.

The guest must reach the airport in 35 minutes. Which route ensures the greater probability of being there on time? Assume that the travel time needed to reach the airport is a random variable with the distribution $\mathsf{N}(\mu_1, \sigma_1^2)$ if route 1 is used and $\mathsf{N}(\mu_2, \sigma_2^2)$ if route 2 is used. Of course, the parameters μ_1, μ_2, σ_1^2 a σ_2^2 are unknown. But the data introduced in Table 9.1 can be considered as independent samples from the mentioned distributions.

An analogous problem is introduced in the book by Wang (1993), who writes that it is a modern version of *Fisher's puzzle* published in the book by Fisher (1934, pp. 123–124) (after Sir Ronald Fisher, 1890–1962). Our solution is based on a method different from Wang's solution.

Consider a general problem. Assume that we have at our disposal data $X_1, \ldots, X_n$ from a given route and the data are a sample from the distribution $\mathsf{N}(\mu, \sigma^2)$. Our further results strongly depend on this assumption of normality. Travel time is influenced by many factors, including vehicular traffic, by waiting times at many crossings, which can be considered as a base for application of the central limit theorem. However, it would be better to have more data and to use a statistical test for normality.

Define

$$\bar{X} = \frac{1}{n}\sum_{i=1}^{n} X_i, \qquad S^2 = \frac{1}{n-1}\sum_{i=1}^{n}(X_i - \bar{X})^2, \qquad Q^2 = \sum_{i=1}^{n}(X_i - \bar{X})^2.$$

The probability π, that travel time X_{n+1} of the next trip to the airport will exceed a number c, is

$$\pi = \mathsf{P}(X_{n+1} > c) = \mathsf{P}\left(\frac{X_{n+1} - \mu}{\sigma} > \frac{c - \mu}{\sigma}\right) = 1 - \Phi\left(\frac{c - \mu}{\sigma}\right),$$

where Φ is the distribution function of $N(0, 1)$. The parameters μ and σ^2 are usually estimated by $\bar{X}$ and S^2. These estimates are unbiased (they satisfy $\mathsf{E}\bar{X} = \mu$ and $\mathsf{E}S^2 = \sigma^2$). The estimate $\bar{X}$ is the *best estimate* for μ (it has the smallest variance among all unbiased estimates). Inserting $\bar{X}$ and S for μ and σ, respectively, we get the *naive estimate* π_N of the probability π in the form

$$\pi_N = 1 - \Phi\left(\frac{c - \bar{X}}{S}\right).$$

Table 9.2 Statistical results.

Route	$\bar{X}$	S^2	Q	π_N	π_K
1	34.5	0.167	0.707	0.110	0.092
2	29.5	49.667	12.207	0.218	0.240

Statisticians do no use the estimate π_N. For example, this estimate is not unbiased (the equality $\mathsf{E}\pi_N = \pi$ does not hold). The best unbiased estimate π_K is preferred. This estimate was derived by Kolmogorov (1950); see Rao [1973, Example 2 in Section 5a(IV)]. Kolmogorov's result can be described as follows. Introduce the function

$$f_n(u) = k_n(1-u^2)^{(n-4)/2}, \qquad -1 < u < 1,$$

where the constant k_n is defined by the condition $\int_{-1}^{1} f_n(u)\,\mathrm{d}u = 1$. This gives

$$k_n = \frac{1}{B\left(\frac{1}{2}, \frac{n-2}{2}\right)},$$

where B is the beta function. Define

$$u_0 = \sqrt{\frac{n}{n-1}}\,\frac{c-\bar{X}}{Q}.$$

Then

$$\pi_K = \begin{cases} 1 & \text{for } u_0 \le -1, \\ \int_{u_0}^{1} f_n(u)\,\mathrm{d}u & \text{for } -1 < u_0 < 1, \\ 0 & \text{for } 1 \le u_0. \end{cases}$$

Note that results of this kind are important in reliability theory. Variables $X_1, \ldots, X_n$ represent lifetimes of some products, and one wishes to estimate the probability that the next product will have a lifetime of at least c. Thus a derivation of the formula for π_K is usually included in some advanced courses of mathematical statistics.

Return to our guest, who must be at the airport within 35 minutes. In the case $n = 4$ we have $f_4(u) = k_4$ for $-1 < u < 1$, and condition $\int_{-1}^{1} f_4(u)\,\mathrm{d}u = 1$ yields that $k_4 = \frac{1}{2}$. Thus

$$\pi_K = \begin{cases} 1 & \text{for } u_0 \le -1, \\ \frac{1}{2}(1-u_0) & \text{for } -1 < u_0 < 1, \\ 0 & \text{for } 1 \le u_0. \end{cases}$$

Results based on data from Table 9.1 are introduced in Table 9.2.

Estimates π_N and π_K do not differ very much. If we take π_K as an estimate of the probability π, then we can say that by using route 1, the guest misses the plane with a probability ~ 0.09, whereas using route 2, this probability is ~ 0.24. This result can be surprising, since all the travel times in route 2 are shorter than all travel times in route 1. The larger probability of missing, when using route 2, is due to the fact that data from this route have a larger variance. Then the next travel can have a larger deviation toward longer travel time.

I was not able to find a similar problem in mentioned Fisher's book. It is really a puzzle.

The preceding story could concern absent-minded mathematicians as they are described in novels and films. However, we all know that they are very rare exceptions. Most of them (or most of us) are very careful, reliable, and conscientious. Consider such a solicitous guest who intends to go from the university to the airport and asks how long time the trip to the airport will take. The next procedure was recommended by J. Machek in a personal communication.

If travel time X to the airport is a random variable with the distribution $\mathsf{N}(\mu, \sigma^2)$, the probability that the travel time is shorter than a given value c is

$$\mathsf{P}(X < c) = \mathsf{P}\left(\frac{X-\mu}{\sigma} < \frac{c-\mu}{\sigma}\right) = \Phi\left(\frac{c-\mu}{\sigma}\right).$$

If we want to ensure that the travel time will be shorter than c with probability p, then c must be a solution of the equation

$$\Phi\left(\frac{c-\mu}{\sigma}\right) = p.$$

Let Φ^{-1} be the quantile function of the distribution $\mathsf{N}(0,1)$, that is, the inverse function of Φ. Then we get

$$c = \mu + \sigma\Phi^{-1}(p).$$

Choose $p = 0.95$, which is the usual level. Then $\Phi^{-1}(0.95) = 1.645$.

Assume that we have independent data $X_1, \ldots, X_n$ for the given route. If we insert estimates $\bar{X}$ and S for corresponding parameters μ and σ, we obtain a *naive estimate* $\hat{c}_p$ for the number c. Hence

$$\hat{c}_p = \bar{X} + S\Phi^{-1}(p).$$

If $p = 0.95$, we have $\hat{c}_{0.95} = 35.17$ for route 1 and $\hat{c}_{0.95} = 41.09$ for route 2. It is a naive estimate of travel time that will not be longer with probability 0.95.

This naive estimate cannot be recommended if the number of observations is small. We substituted unknown parameters by their estimates, and this could lead to considerable inaccuracy. We describe another procedure that is used in technical applications. Instead of the number $\Phi^{-1}(p)$ which occurs in the formula for $\hat{c}_p$, we use another constant k. We choose k such that a new random variable

$$\tilde{c}_p = \bar{X} + kS$$

satisfies

$$\mathsf{P}[\tilde{c}_p > \mu + \Phi^{-1}(p)\sigma] = \gamma,$$

where γ is a sufficiently large probability. Let $\gamma = 0.95$, for simplicity. There is a large probability that the variable $\tilde{c}_p$ is not smaller than the time which will not be exceeded in the trip to the airport with probability p. We introduce random variables

$$U = \frac{\bar{X} - \mu}{\sigma}\sqrt{n}, \qquad Y = \frac{(n-1)S^2}{\sigma^2}, \qquad \xi = \frac{U - \Phi^{-1}(p)\sqrt{n}}{\sqrt{\frac{Y}{n-1}}}.$$

After some computation we get

$$\begin{aligned}
\gamma &= \mathsf{P}[\bar{X} + kS > \mu + \Phi^{-1}(p)\sigma] = \mathsf{P}[\bar{X} - \mu - \sigma\Phi^{-1}(p) > -kS] \\
&= \mathsf{P}\left(\frac{\frac{\bar{X}-\mu}{\sigma}\sqrt{n} - \Phi^{-1}(p)\sqrt{n}}{\frac{S}{\sigma}} > -k\sqrt{n}\right) \\
&= \mathsf{P}(\xi > -k\sqrt{n}).
\end{aligned}$$

It is known that $U \sim \mathsf{N}(0,1), Y \sim \chi^2_{n-1}$, and that U and Y are mutually independent. Then the random variable ξ has a noncentral Student distribution $\mathsf{t}_{n-1,\delta}$ with $n-1$ degrees of freedom and a parameter of noncentrality $\delta = -\Phi^{-1}(p)\sqrt{n}$. Let $\tau_{n-1,\delta}$ be the quantile function of the distribution $\mathsf{t}_{n-1,\delta}$. (Remember that $\tau_{n-1,\delta}$ is the inverse function of the distribution function of the distribution $\mathsf{t}_{n-1,\delta}$.) From the condition $\gamma = \mathsf{P}(\xi > -k\sqrt{n}) = 1 - \mathsf{P}(\xi \le -k\sqrt{n})$ derived above, it follows that $\mathsf{P}(\xi \le -k\sqrt{n}) = 1 - \gamma$, so that $-k\sqrt{n} = \tau_{n-1,\delta}(1-\gamma)$. This gives

$$k = -\frac{\tau_{n-1,\delta}(1-\gamma)}{\sqrt{n}}.$$

In our case we obtain the following numerical values:

$$n = 4, \quad \Phi^{-1}(0.95) = 1.645, \quad \delta = -3.290, \quad \tau_{3,\delta}(0.95) = -10.280.$$

In the case of route 1 we have $\tilde{c}_{0.95} = 36.59$; for route 2, we get $\tilde{c}_{0.95} = 65.54$. The results indicate that the guest should reserve time about 37 minutes for route 1 and about 1 hour and 6 minutes for route 2. The difference between our result and the naive estimate is rather large, especially for route 2.

9.5 CHRISTMAS INEQUALITY

Consider a sequence of independent random variables $X_1, X_2, \ldots$ with expectation μ and a finite variance $\sigma^2 > 0$. Define

$$S_n = X_1 + X_2 + \cdots + X_n.$$

It is clear that

$$\mathsf{E}S_n = n\mu, \qquad \operatorname{var} S_n = n\sigma^2.$$

Thus $\mathsf{var}\, S_n$ depends on n linearly. Now assume additionally that the variables X_i are nonnegative. We can ask how $\mathsf{var}\,\sqrt{S_n}$ depends on n. One could expect that the variance grows like $\sqrt{n}$, but we shall see that this is not so. Behavior of the expression $\mathsf{var}\, S_n^{1/4}$ seems to be even more interesting, and we shall consider it later. Such problems were investigated in papers by Banjevič and Bratičevič (1983) and Banjevič (1991).

First, we notice that assumptions $X_i \geq 0$ and $\sigma^2 > 0$ imply that $\mu > 0$. Further, we introduce some elementary assertions.

Theorem 9.2 *Let Y be a random variable such that $\mathsf{E}Y^2 < \infty$. Then we have for arbitrary real a that*

$$\mathsf{E}(Y - \mathsf{E}Y)^2 \leq \mathsf{E}(Y - a)^2.$$

Proof. We can write

$$\begin{aligned} \mathsf{E}(Y - a)^2 &= \mathsf{E}[(Y - \mathsf{E}Y) + (\mathsf{E}Y - a)]^2 \\ &= \mathsf{E}(Y - \mathsf{E}Y)^2 + (\mathsf{E}Y - a)^2 \geq \mathsf{E}(Y - \mathsf{E}Y)^2. \quad \square \end{aligned}$$

Theorem 9.3 *Let $a \geq 0$, $b > 0$. Then*

$$(a - b)^2 \leq \frac{(a^2 - b^2)^2}{b^2}.$$

Proof. Since $(a + b)^2/b^2 \geq 1$, we have

$$(a - b)^2 \leq (a - b)^2 \frac{(a + b)^2}{b^2} = \frac{(a^2 - b^2)^2}{b^2}. \quad \square$$

Theorem 9.4 *Let S be a nonnegative random variable with a positive variance. Then*

$$\mathsf{var}\,\sqrt{S} \leq \frac{\mathsf{var}\, S}{\mathsf{E}S}.$$

Proof. Applying Theorems 9.2 and 9.3, we get

$$\begin{aligned} \mathsf{var}\,\sqrt{S} &= \mathsf{E}(\sqrt{S} - \mathsf{E}\sqrt{S})^2 \leq \mathsf{E}(\sqrt{S} - \sqrt{\mathsf{E}S})^2 \\ &\leq \frac{\mathsf{E}(S - \mathsf{E}S)^2}{\mathsf{E}S} = \frac{\mathsf{var}\, S}{\mathsf{E}S}. \quad \square \end{aligned}$$

From this result we easily obtain the answer to the first question. If we put $S = S_n$, then

$$\mathsf{var}\,\sqrt{S_n} \leq \frac{\sigma^2}{\mu}. \tag{9.2}$$

The inequality (9.2) was called the *"Christmas inequality"*, since some papers on this formula were published in December issues if the *Information Bulletin of the Czech Statistical Society* (Anděl 1990, 1991a, 1991b).

Before investigating $\mathsf{var}\, S_n^r$ for a general exponent $r \in [0, 1]$, we generalize Theorem 9.3.

Theorem 9.5 *Let $a \geq 0$, $b > 0$, $r \in [0, 1]$. Then*

$$(a^r - b^r)^2 \leq \frac{(a-b)^2}{b^{2(1-r)}}.$$

Proof. If $t \geq 0$, then $|t^r - 1| \leq |t - 1|$. Thus $(t^r - 1)^2 \leq (t-1)^2$. Let $t = a/b$. Then

$$\left[\left(\frac{a}{b}\right)^r - 1\right]^2 \leq \left(\frac{a}{b} - 1\right)^2,$$

and thus

$$(a^r - b^r)^2 b^{2(1-r)} \leq (a-b)^2. \quad \square$$

Theorem 9.6 *Let S be a nonnegative random variable with a positive variance. Assume that $r \in [0, 1]$. Then*

$$\operatorname{var} S^r \leq \frac{\operatorname{var} S}{(\mathsf{E}S)^{2(1-r)}}.$$

Proof. Theorems 9.2 and 9.5 yield

$$\begin{aligned} \operatorname{var} S^r &= \mathsf{E}(S^r - \mathsf{E}S^r)^2 \leq \mathsf{E}[S^r - (\mathsf{E}S)^r]^2 \\ &\leq \frac{\mathsf{E}(S - \mathsf{E}S)^2}{(\mathsf{E}S)^{2(1-r)}} = \frac{\operatorname{var} S}{(\mathsf{E}S)^{2(1-r)}}. \quad \square \end{aligned}$$

If we again put $S = S_n$, we obtain

$$\operatorname{var} S_n^r \leq \frac{n\sigma^2}{(n\mu)^{2(1-r)}} = n^{-1+2r} \frac{\sigma^2}{\mu^{2(1-r)}}.$$

For $r \in [0, \frac{1}{2})$, we obtain $\operatorname{var} S_n^r \to 0$ as $n \to \infty$. In particular, $\operatorname{var} S_n^{1/4} \to 0$ as $n \to \infty$.

9.6 CINDERELLA

In the well-known fairytale, *Cinderella*, returning home from the ball, lost her little shoe. The prince noticed that the shoe was extraordinary small. Royal messengers successively tried the little shoe on other young girls. When they found the first girl whom the shoe fit, they cheered that it was Cinderella.

Assume that N girls of the corresponding age live in the kingdom and M of them can put on the small shoe. Let $M < N$. Assume that the royal messengers try the shoe on girls in random order.

Let X be the random variable that is equal to the number of failures until the first success. It is clear that we have the following for $k = 0, 1, \ldots, N - M$:

$$\begin{aligned} \mathsf{P}(X = k) &= \frac{N-M}{N}\frac{N-M-1}{N-1}\cdots\frac{N-M-k+1}{N-k+1}\frac{M}{N-k} \\ &= \frac{\binom{N-k-1}{M-1}}{\binom{N}{M}}. \end{aligned}$$

We denote this distribution by $\mathsf{D}(N, M)$. Of course, we have

$$\mathsf{P}(X = 0) + \mathsf{P}(X = 1) + \cdots + \mathsf{P}(X = N - M) = 1.$$

Calculation of $\mathsf{E}X$ is based on the formula

$$\mathsf{E}(N - M - X) = \sum_{k=0}^{N-M} (N - M - k)\mathsf{P}(X = k).$$

The last term of this expression vanishes; we can omit it, and we get

$$\begin{aligned} \mathsf{E}(N - M - X) &= \sum_{k=0}^{N-M-1} \frac{N-M}{N}\cdots\frac{N-M-k+1}{N-k+1}\frac{M}{N-k} \\ &= \frac{(N-M)M}{M+1}\sum_{k=0}^{N-(M+1)} \frac{N-(M+1)}{N}\cdots\frac{N-(M+1)-k+1}{N-k+1}\frac{M+1}{N-k} \\ &= \frac{(N-M)M}{M+1}, \end{aligned}$$

since the last sum is 1 because it is the sum of all probabilities of the distribution $\mathsf{D}(N, M + 1)$. Thus we obtain

$$\mathsf{E}X = N - M - \frac{(N-M)M}{M+1} = \frac{N-M}{M+1}.$$

The probability that this method discovers Cinderella is $1/M$, since random order ensures that any girl with a small foot can be visited by the royal messengers as the first.

Royal statisticians have a good idea about the number N, and so we shall assume that this number is known. If they cooperate with the shoeindustry, they can have a good estimate of the number M. In such a case they must know that (in the Czech Republic) the average length of the sole of girls of age 17 is 242 mm, with a standard deviation of 12 mm. Since N is known, this information (and some others, too) gives quite a good estimate of M. If the only information on M is that $1 \le M < N$ and nothing else, royal statisticians could use knowledge of a number of unsuccessful trials before the first successful one.

The simplest procedure can be based on the *moment method*. Here the expectation of random variable X is taken to be equal to its realized value k. This leads to the equation

$$\frac{N-M}{M+1} = k,$$

the solution $M = M^*$ of which is

$$M^* = \frac{N-k}{k+1}.$$

This result must be rounded to the nearest integer.

Another possibility is to use the *maximum likelihood method*. The estimate of the parameter M is a number $\hat{M}$ such that probability $\mathsf{P}(X = k)$ is maximal for given N and k. Since the probability $\mathsf{P}(X = k)$ is considered as a function of variable M in this case, we denote it as $L(M)$. Since

$$\frac{L(M)}{L(M-1)} = \frac{N-M-k+1}{N-M+1}\frac{M}{M-1},$$

the relation $L(M)/L(M-1) \leq 1$ holds if and only if

$$M \leq \frac{N+1}{k+1}.$$

Similarly, from the formula

$$\frac{L(M+1)}{L(M)} = \frac{N-M-k}{N-M}\frac{M+1}{M}$$

we find that $L(M+1)/L(M) \leq 1$ holds if and only if

$$\frac{N-k}{k+1} \leq M.$$

This implies that $L(M)$ is maximal for $M = \hat{M}$ such that the inequalities

$$\frac{N-k}{k+1} \leq \hat{M} \leq \frac{N+1}{k+1}$$

hold. Since the difference between right and left bounds is exactly 1, there exist either one or two integer-valued $\hat{M}$ satisfying the inequalities. We can see that the estimate based on the moment method equals the left-hand bound for $\hat{M}$.

Only in the case $k = N - 1$ are we sure the first found girl with the small foot is Cinderella. On the other hand, our estimate M^* is equal to 1 even if $k = (N-1)/2$.

Some other information on statistical aspects of the fairytale Cinderella can be found in Komenda (1997). We investigate the behavior of the estimate M^* and calculate $\mathsf{E}M^*$. Our calculation is based on the formula

$$\sum_{i=k}^{n} \binom{i}{k} \frac{1}{n+1-i} = \binom{n+1}{k} \sum_{i=k}^{n} \frac{1}{i+1}, \qquad 0 \leq k \leq n,$$

Table 9.3 Approximation of values $\mathsf{E}M^*$ for $M = 1$.

N	10	100	1000	10,000	100,000	1,000,000
Approximation $\mathsf{E}M^*$	2.2	4.2	6.5	8.8	11.1	13.4

which was published as Problem E2084 in *Am. Math. Monthly* (**76**, 1969, p. 420; see Kaucký 1975, p. 69, Problem 48). We successively obtain

$$\begin{aligned}
\mathsf{E}M^* &= \sum_{k=0}^{N-M} \frac{N-k}{k+1} \frac{\binom{N-k-1}{M-1}}{\binom{N}{M}} \\
&= \frac{M}{\binom{N}{M}} \sum_{k=0}^{N-M} \frac{1}{k+1} \binom{N-k}{M} \\
&= \frac{M}{\binom{N}{M}} \sum_{j=0}^{N-M} \frac{1}{N-M-j+1} \binom{M+j}{M} \\
&= \frac{M}{\binom{N}{M}} \sum_{i=M}^{N} \frac{1}{N+1-i} \binom{i}{M} \\
&= \frac{M}{\binom{N}{M}} \binom{N+1}{M} \sum_{i=M}^{N} \frac{1}{i+1} \\
&= M \frac{N+1}{N-M+1} \left(\frac{1}{M+1} + \frac{1}{M+2} + \cdots + \frac{1}{N+1} \right).
\end{aligned}$$

For example, in case $M = 1$ using the Euler formula (4.4), we have

$$\mathsf{E}M^* \doteq \frac{N+1}{N}[\ln(N+1) + \gamma - 1].$$

These values for different N are introduced in Table 9.3.

Using (4.4), for larger values of M, we would obtain

$$\mathsf{E}M^* \doteq M \frac{N+1}{N-M+1} \ln \frac{N+1}{M}.$$

If N is large and $M = 0.03N$, we get $\mathsf{E}M^* \doteq 3.6M$. Obviously the estimate M^* is not unbiased, but it overestimates the true value M.

10

The LAD method

10.1 MEDIAN

Consider a straight road with a few houses on one side. Somewhere among these houses, construction of a telephone center is planned. For some reason this center should be connected with each house by a wire. Find the place for the telephone center such that the total length of the wire is minimal.

We construct a model to illustrate this situation. Instead of the road, we take a straight line and represent the houses by their coordinates $x_1, \ldots, x_n$ with respect to an arbitrary origin at the line. If the telephone center is placed at a point a, the length of the wire connecting the center and the ith point is $|x_i - a|$. Hence the total length of the wire is

$$f(a) = \sum_{i=1}^{n} |x_i - a|. \tag{10.1}$$

One would think that f reaches its minimum at the point $a = \bar{x}$, where

$$\bar{x} = \frac{1}{n} \sum_{i=1}^{n} x_i$$

is the average. However, this is not the case.

Order all numbers $x_1, \ldots, x_n$ into a nondecreasing sequence. Let $x_{(1)}$ be the smallest one, $x_{(2)}$ the second smallest one, and so on. This means that $x_{(1)} \leq x_{(2)} \leq \cdots \leq x_{(n)}$. If n is odd, say, $n = 2m - 1$, we define the *median* $\tilde{x}$ as the middle value, specifically, $\tilde{x} = x_{(m)}$. If n is even, say, $n = 2m$, then the median is an arbitrary number between the middle two terms. In this case $\tilde{x} \in [x_{(m)}, x_{(m+1)}]$. We can see

Table 10.1 Data for fitting a line.

i	1	2	3	4	5
x_i	1	2	3	4	5
y_i	1	4	2	6	4

that in the general case the median is not determined uniquely. For this reason, the definition $\tilde{x} = (x_{(m)} + x_{(m+1)})/2$ can be found for $n = 2m$ in some textbooks. We do not use this specification in this book.

We prove that the problem of the planned site for the telephone center is solved by a median.

Theorem 10.1 *Function f introduced in (10.1) reaches its minimum at $a = \tilde{x}$.*

Proof. Assume first that all numbers $x_1, \ldots, x_n$ are different. Obviously f is a continuous function of the variable a. If $a < x_{(1)}$, then

$$f(a) = \sum_{i=1}^{n} (x_i - a) = n(\bar{x} - a),$$

so that $f'(a) = -n$. Similarly, for $a > x_{(n)}$, we have $f'(a) = n$. Let $a \in (x_{(j)}, x_{(j+1)})$ for some $j = 1, \ldots, n-1$. Then

$$f(a) = \sum_{i=1}^{j} |x_{(i)} - a| + \sum_{i=j+1}^{n} |x_{(i)} - a| = -\sum_{i=1}^{j} x_{(i)} + ja + \sum_{i=j+1}^{n} x_{(i)} - (n-j)a,$$

and thus $f'(a) = (2j-n)$. The function $f(a)$ decreases for $j < n/2$ and increases for $j > n/2$. If n is an even number, then $f(a)$ is constant in the interval $\left[x_{(\frac{n}{2})}, x_{(\frac{n}{2}+1)}\right]$. If some of the numbers x_i are equal, the proof is similar. □

10.2 LEAST SQUARES METHOD

In many cases there is proven linear interdependence between two variables. Remember elementary version of Hooke's law (after Robert Hooke, 1635–1703) in physics, which describes the extension of a stick under loading. But the data actually measured do not lie on a line exactly because of some small errors during the measurement. If we plot the data on a graph, it is a natural problem to find a straight line such that it fits the points as closely as possible. For example, points $(x_1, y_1), \ldots, (x_5, y_5)$ introduced in Table 10.1 are depicted in Fig. 10.1.

Assume that the points $(x_1, y_1), \ldots (x_n, y_n)$ are given. Let $n \geq 2$. We exclude the case that $x_1 = \cdots = x_n$. Adrien-Marie Legendre (1752–1833) and Gauss proposed drawing a segment from each point to the line in question in a vertical direction. This

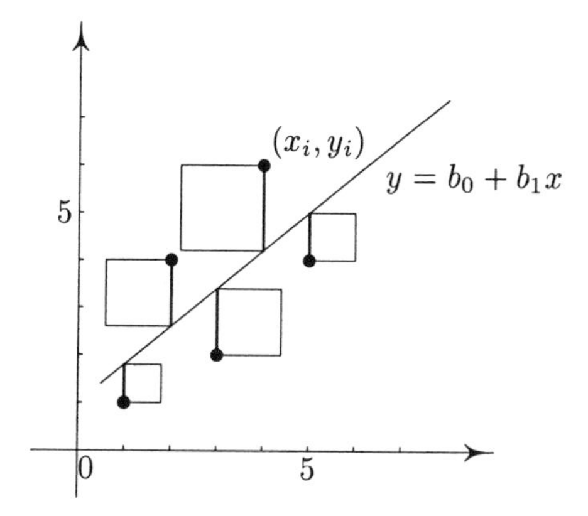

Fig. 10.1 Least squares method.

segment is taken as a side of a square. The best fitted line is the one that has the minimal sum of squares. We say that the straight line is obtained by the *least squares* (LS) *method*.

The side of the ith square is $|y_i - b_0 - b_1 x_i|$. The sum of squares is a function of the parameters b_0 and b_1. We denote this by

$$S(b_0, b_1) = \sum_{i=1}^{n} (y_i - b_0 - b_1 x_i)^2.$$

The values b_0 and b_1 minimizing the function f can be calculated using partial derivatives. Since

$$\begin{aligned} \frac{\partial S}{\partial b_0} &= -2 \sum (y_i - b_0 - b_1 x_i), \\ \frac{\partial S}{\partial b_1} &= -2 \sum x_i (y_i - b_0 - b_1 x_i), \end{aligned}$$

from the conditions

$$\frac{\partial S}{\partial b_0} = 0, \qquad \frac{\partial S}{\partial b_1} = 0,$$

we get equations

$$\begin{aligned} n b_0 + \left(\sum x_i\right) b_1 &= \sum y_i, \\ \left(\sum x_i\right) b_0 + \left(\sum x_i^2\right) b_1 &= \sum x_i y_i. \end{aligned} \tag{10.2}$$

From this system b_0 and b_1 can be easily calculated. It can be checked in the standard way that this solution gives a minimum of the function f. By the way, the problem is so simple that it can be solved without derivatives (see Anděl 1985/86).

In our example (see Table 10.1 and Fig. 10.1) we have

$$n = 5, \quad \sum x_i = 15, \quad \sum x_i^2 = 55,$$
$$\sum y_i = 17, \quad \sum x_i y_i = 59.$$

Inserting terms into (10.2), we obtain

$$\begin{aligned} 5b_0 + 15b_1 &= 17, \\ 15b_0 + 55b_1 &= 59, \end{aligned}$$

the solution of which is $b_0 = 1$, $b_1 = 0.8$. This straight line is drawn in Fig. 10.1. It has the *residual sum of squares* $S(b_0, b_1) = 8.8$.

Sometimes the straight line must go through the origin. Its equation is $y = bx$. The least squares method leads to the condition that the function

$$S(b) = \sum (y_i - bx_i)^2$$

must be minimal. If at least one of the values $x_1, \ldots, x_n$ is nonvanishing, it can be easily checked that the minimum is reached at the point

$$b = \frac{\sum x_i y_i}{\sum x_i^2}.$$

Using data introduced in Table 10.1, we get $b = 1.07273$. In this case the residual sum of squares is $S(b) = 9.709$. Of course, this can never be smaller than the residual sum of squares corresponding to the general straight line — it was 8.8.

10.3 LAD METHOD

The least squares method is very popular, since it gives simple explicit formulas. However, each deviation from the straight line is represented in the function S by its square, and so the method is very sensitive to gross errors and outliers. A natural alternative criterion seems to be to minimize only the sum of the lengths of the segments that are drawn from each point in a vertical direction to the straight line. Then in the case of a general straight line we minimize the function

$$L(b_0, b_1) = \sum_{i=1}^{n} |y_i - b_0 - b_1 x_i|,$$

whereas in the case of the straight line going through the origin, we minimize

$$L(b) = \sum_{i=1}^{n} |y_i - bx_i|.$$

This is called the *least absolute deviations* (LAD) *method*. We show that the problem always has a solution but, in general, the solution is not unique. It is not difficult to minimize the function $L(b)$. It was Laplace who found the solution, and we present his procedure in the next section. Application of the LAD method in the case of a general straight line is more complicated, and we discuss the problem in Section 10.5.

10.4 LAPLACE METHOD

When minimizing function $L(b)$ without loss of generality, we can restrict our discussion to the points at which $x_i \neq 0$. Specifically, we have

$$L(b) = \sum_{\{i:x_i=0\}} |y_i| + \sum_{\{i:x_i \neq 0\}} |y_i - bx_i|.$$

Function $L(b)$ is minimized by the values b such that they minimize the second sum on the right-hand side.

In this section we assume that the points (x_i, y_i) are ordered in such a way that

$$\frac{y_i}{x_i} \leq \frac{y_{i+1}}{x_{i+1}}.$$

Theorem 10.2 *There exists $x_i \neq 0$ such that $b = y_i/x_i$ minimizes the function L. One such index i is given by the formula*

$$i = \min\left\{j : \sum_{k=1}^{j} |x_k| \geq \frac{1}{2}\sum_{k=1}^{n} |x_k|\right\}.$$

Proof. Let

$$b \in \left(\frac{y_p}{x_p}, \frac{y_{p+1}}{x_{p+1}}\right).$$

Then

$$L(b) = \sum_{i=1}^{p} |x_i| \left(b - \frac{y_i}{x_i}\right) - \sum_{i=p+1}^{n} |x_i| \left(b - \frac{y_i}{x_i}\right).$$

The derivative is

$$L'(b) = \sum_{i=1}^{p} |x_i| - \sum_{i=p+1}^{n} |x_i|.$$

Thus $L'(b)$ is nondecreasing. Since L is continuous, it consists of linear parts and its derivative is nondecreasing. Its minimum is at the smallest of the points such that the derivative there is nonnegative. Obviously, such an index is given by the formula

$$i = \min\left\{j : \sum_{k=1}^{j} |x_k| \geq \sum_{k=j+1}^{n} |x_k|\right\}.$$

This expression is equivalent to the formula introduced in Theorem 10.2. □

We apply the *Laplace method* to the data introduced in Table 10.1. We order the points according to increasing values of the ratio y_i/x_i. The results are given in Table 10.2. The values $\sum_{k=1}^{j} |x_k|$ can be found in Table 10.3.

We have $\frac{1}{2}\sum_{k=1}^{5} |x_k| = 7.5$. We can see from Table 10.3 that the value $\sum_{k=1}^{j} |x_k|$ is for the first time equal to or greater than 7.5 for $j = 2$. Thus the calculated straight

Table 10.2 Application of Laplace method.

i	1	2	3	4	5
x_i	3	5	1	4	2
y_i	2	4	1	6	4
y_i/x_i	$\frac{2}{3}$	$\frac{4}{5}$	1	$\frac{3}{2}$	2

Table 10.3 Values $\sum_{k=1}^{j} |x_k|$.

j	1	2	3	4	5
$\sum_{i=1}^{j} \lvert x_i \rvert$	3	8	9	13	15

line goes through the point (5,4) and its equation is $y = 0.8x$. If $b = 0.8$, then $L(b) = 5.8$. It may be interesting to compare this result with the value of the function L at the point 1.07273, which is the parameter of the straight line going through the origin using the least squares method. We get $L(1.072\,73) = 6.218\,19$. It is a little surprizing that for the general straight line calculated by the least squares method we obtain even a larger number $\sum_{k=1}^{5} |y_k - 1 - 0.8x_k| = 6.4$.

10.5 GENERAL STRAIGHT LINE

For given points $(x_1, y_1), \ldots, (x_n, y_n)$, calculate coefficients b_0, b_1 of the straight line $y = b_0 + b_1 x$ such that the function

$$L(b_0, b_1) = \sum_{i=1}^{n} |y_i - b_0 - b_1 x_i|$$

is minimal. Assume that at least two numbers among $x_1, \ldots, x_n$ are different. Gauss proved (see Theorem 10.3 later, in Section 10.6) that there exists an optimal straight line such that it goes through at least two points: (x_i, y_i) and $(x_j, y_j), i \neq j$. Were all given points different, it would suffice to analyze maximally $\binom{n}{2}$ straight lines, which are determined by these points. The value $L(b_0, b_1)$ would be calculated for each line, and we would choose the line with minimal $L(b_0, b_1)$. In practical calculations more economical methods are preferred. It follows from Theorem 10.4 that minimization of function $L(b_0, b_1)$ is equivalent to the problem involving minimization of $\sum_{i=1}^{n} r_i$ under the constraints

$$\begin{aligned} r_i + b_0 + b_1 x_i &\geq \ \ y_i, \qquad i = 1, \ldots, n, \\ r_i - b_0 - b_1 x_i &\geq -y_i, \qquad i = 1, \ldots, n. \end{aligned}$$

This is a linear programming problem with $n + 2$ variables $r_1, \ldots, r_n, b_0, b_1$, and $2n$ constraints. In the example with data from Table 10.1, this problem can be written in the form

$$\text{Minimize } \boldsymbol{c}'\boldsymbol{\xi} = \sum_{i=1}^{5} c_i \xi_i \quad \text{under constraints } \boldsymbol{A}\,\boldsymbol{\xi} \geq \boldsymbol{b},$$

where

$$\boldsymbol{A} = \begin{pmatrix} 1 & 0 & 0 & 0 & 0 & 1 & x_1 \\ 0 & 1 & 0 & 0 & 0 & 1 & x_2 \\ 0 & 0 & 1 & 0 & 0 & 1 & x_3 \\ 0 & 0 & 0 & 1 & 0 & 1 & x_4 \\ 0 & 0 & 0 & 0 & 1 & 1 & x_5 \\ 1 & 0 & 0 & 0 & 0 & -1 & -x_1 \\ 0 & 1 & 0 & 0 & 0 & -1 & -x_2 \\ 0 & 0 & 1 & 0 & 0 & -1 & -x_3 \\ 0 & 0 & 0 & 1 & 0 & -1 & -x_4 \\ 0 & 0 & 0 & 0 & 1 & -1 & -x_5 \end{pmatrix}, \quad \boldsymbol{b} = \begin{pmatrix} y_1 \\ y_2 \\ y_3 \\ y_4 \\ y_5 \\ -y_1 \\ -y_2 \\ -y_3 \\ -y_4 \\ -y_5 \end{pmatrix}, \quad \boldsymbol{\xi} = \begin{pmatrix} r_1 \\ r_2 \\ r_3 \\ r_4 \\ r_5 \\ b_0 \\ b_1 \end{pmatrix},$$

and $\boldsymbol{c} = (1,1,1,1,1,0,0)'$. A computer gives the solution $r_1 = 0$, $r_2 = 2.25$, $r_3 = 0.5, r_4 = 2.75, r_5 = 0, b_0 = 0.25, b_1 = 0.75$.

The value r_i is the length of the segment going from the point (x_i, y_i) in a vertical direction to the optimal straight line $y = b_0 + b_1 x$. Since $r_1 = 0$, $r_5 = 0$, this means that the line goes through the first and fifth point; thus the points $(1,1)$ and $(5,4)$ lie on the line. This is in accordance with the Gauss theorem. The equation of the optimal line is $y = 0.25 + 0.75x$ and the computer also gives the value of the objective function

$$L(b_0, b_1) = \sum_{i=1}^{5} r_i = 5.5.$$

The number $L(b_0, b_1)$ cannot be larger than the optimal value $L(b)$ for the line going through the origin. In our case we got $L(b) = 6.21819$.

10.6 LAD METHOD IN A GENERAL CASE

In a general case vectors $(x_{i1}, \ldots, x_{ik}, y_i)$ for $i = 1, \ldots, n$ are given and the problem is to find $\boldsymbol{\beta} = (\beta_1, \ldots, \beta_k)$ such that the function

$$L(\boldsymbol{\beta}) = \sum_{i=1}^{n} \left| y_i - \sum_{j=1}^{k} x_{ij}\beta_j \right|$$

is minimal. Gauss derived the following theorem in 1809.

Theorem 10.3 (Gauss) *Let $\boldsymbol{X} = (x_{ij})$ be a matrix of the type $n \times k$ with elements x_{ij}. Assume that the rank of the matrix $\boldsymbol{X}$ is $h \geq 1$. Then there exists a vector $\boldsymbol{\beta}$*

minimizing the function L such that at least h differences

$$\Delta_i = y_i - \sum_{j=1}^{k} x_{ij}\beta_j$$

vanish.

Proof. Define $\boldsymbol{x}_i = (x_{i1}, \ldots, x_{ik})'$. Let $Z_{\boldsymbol{\beta}} = \{i : y_i = \boldsymbol{\beta}'\boldsymbol{x}_i\}$. Let p be the number of elements of the set $Z_{\boldsymbol{\beta}}$. Assume that $p < h$. Then there exists $\mathbf{d} = (d_1, \ldots, d_k)' \neq \mathbf{0}$ such that $\mathbf{d}'\boldsymbol{x}_i = 0$ for $i \in Z_{\boldsymbol{\beta}}$ and $\mathbf{d}'\boldsymbol{x}_i \neq 0$ for at least one $i \notin Z_{\boldsymbol{\beta}}$. Let t be a real number. Denote

$$K(t) = \sum_{i=1}^{n} |y_i - \boldsymbol{\beta}\boldsymbol{x}_i - t\mathbf{d}'\boldsymbol{x}_i| = \sum_{i \notin Z_{\boldsymbol{\beta}}} |y_i - \boldsymbol{\beta}\boldsymbol{x}_i - t\mathbf{d}'\boldsymbol{x}_i|.$$

If we write $w_i = y_i - \mathbf{c}'\boldsymbol{x}_i$, $z_i = \mathbf{d}'\boldsymbol{x}_i$, then

$$K(t) = \sum_{i \notin Z_{\boldsymbol{\beta}}} |w_i - tz_i|.$$

Theorem 10.2 implies that there exists $q \notin Z_{\boldsymbol{\beta}}$ such that $t = t_q = w_q/z_q$ minimizes the function $K(t)$. It is clear that $t_q \neq 0$, since $w_i \neq 0$ for each $i \notin Z_{\boldsymbol{\beta}}$. Then the considered vector $\boldsymbol{\beta}$ satisfies

$$L(\boldsymbol{\beta}) = K(0) \geq K(t_q).$$

The new vector of parameters $\gamma = \boldsymbol{\beta} + t\mathbf{d}$ has the property that the differences $y_i - \boldsymbol{x}_i'\gamma$ vanish not only for $i \in Z_{\boldsymbol{\beta}}$ but also for $i = q$. □

Theorem 10.4 *The problem of minimize function $L(\boldsymbol{\beta})$ is equivalent to the problem of linear programming*

$$\textit{Minimize} \quad \sum_{i=1}^{n} r_i$$

under constraints

$$r_i + \sum_{j=1}^{k} x_{ij}\beta_j \geq y_i, \qquad r_i - \sum_{j=1}^{k} x_{ij}\beta_j \geq -y_i \qquad (i = 1, \ldots, n).$$

Proof. The constraints can be written in the equivalent form

$$r_i \geq y_i - \sum_{j=1}^{k} x_{ij}\beta_j, \qquad r_i \geq \sum_{j=1}^{k} x_{ij}\beta_j - y_i \qquad (i = 1, \ldots, n),$$

which implies that $r_i \geq |y_i - \sum_{j=1}^{k} x_{ij}\beta_j|$.

Let $\boldsymbol{\beta}$ minimize the function L. For arbitrary $r_i \geq |y_i - \boldsymbol{\beta}\boldsymbol{x}_i|$, the values $r_1, \ldots, r_n$, $\boldsymbol{\beta}$ satisfy conditions of the linear programming problem and $\sum r_i \geq L(\boldsymbol{\beta})$. If we take $r_i = |y_i - \boldsymbol{\beta}'\boldsymbol{x}_i|$, then $\sum r_i = L(\boldsymbol{\beta})$. This implies that the solution of the linear programming problem satisfies $r_i \leq L(\boldsymbol{\beta})$. $\square$

A detailed description of the LAD method can be found in Bloomfield and Steiger (1983).

11

Probability in mathematics

11.1 QUADRATIC EQUATIONS

The *quadratic equation* $x^2 + px + q = 0$ has roots

$$x_{12} = \frac{1}{2}\left(-p \pm \sqrt{p^2 - 4q}\right).$$

The roots are *imaginary* if and only if

$$q > \frac{p^2}{4}. \tag{11.1}$$

Let n be a positive number. Assume that p and q are independent random variables that take all integer values from the interval $[-n, n]$ with the same probability. In this case n is usually chosen such that it is also an integer. Then the number of points with integer-valued coordinates in the square $[-n, n] \times [-n, n]$ is $(2n + 1)^2$. Let m_n be the number of integer-valued points from the square that are located over the parabola $q = p^2/4$ (see Figs. 11.1 and 11.2). Then the probability of the event (11.1) is

$$P_n = \frac{m_n}{(2n+1)^2}.$$

We can see in Fig. 11.1 that $m_3 = 15$ and in Fig. 11.2 that $m_5 = 31$. Thus

$$P_3 = \frac{15}{49} = 0.306, \qquad P_5 = \frac{31}{121} = 0.256.$$

Some values m_n and P_n are given in Table 11.1 (see Anděl and Zvára 1980/81).

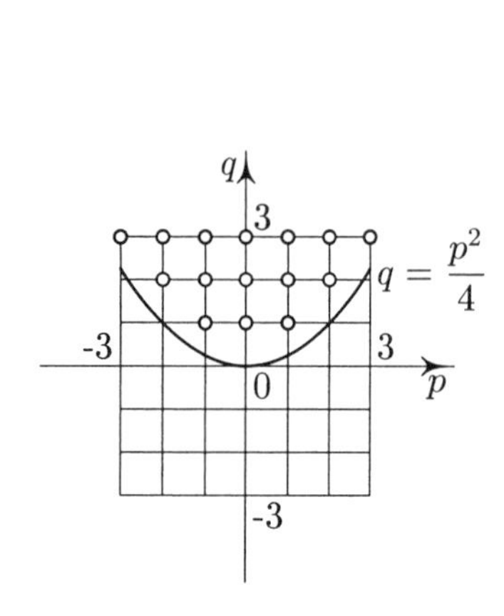

Fig. 11.1 Imaginary roots, $n = 3$.

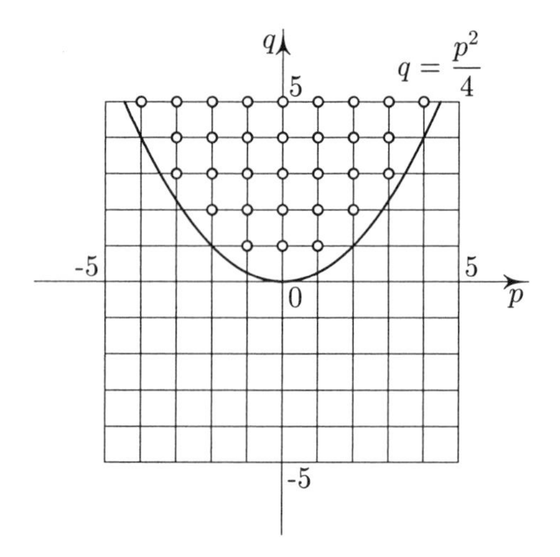

Fig. 11.2 Imaginary roots, $n = 5$.

Consider the case when p and q are independent random variables with continuous rectangular distribution on the interval $[-n, n]$. Here we have no reason to restrict ourselves only to integer n. Probability P_n^* of the event (11.1) is the ratio of the area inside the square $[-n, n] \times [-n, n]$ over the parabola $q = p^2/4$ to the area of the whole square $(2n)^2$. Integration gives

$$P_n^* = \begin{cases} 0.5 - \dfrac{n}{24} & \text{for } 0 < n \leq 4, \\ \dfrac{2}{3\sqrt{n}} & \text{for } n \geq 4. \end{cases}$$

Some values P_n^* are also introduced in Table 11.1. It is remarkable that $P_n^* \to 0$ as $n \to \infty$. Some people have even argued that complex numbers were introduced so late because imaginary roots were not very probable. But this is only speculation.

Consider the general form of the quadratic equation $ax^2 + bx + c = 0$. Let a, b, and c be integer-valued independent random variables, where b and c take with the same probability any integer value from the interval $[-n, n]$ and a takes with the same probability any integer value from the same interval $[-n, n]$ except from zero. Let n be a positive integer. We have $2n(2n + 1)^2$ equally probable triples (a, b, c). Our quadratic equation has imaginary roots in the case that

$$b^2 < 4ac. \tag{11.2}$$

Let M_n be the number of triples (a, b, c) such that the condition (11.2) is satisfied. Then the probability of the event (11.2) is

$$Q_n = \frac{M_n}{2n(2n + 1)^2}.$$

Table 11.1 Probabilities of imaginary roots.

n	5	10	15	20
m_n	31	86	157	240
P_n	0.256	0.195	0.163	0.143
P_n^*	0.298	0.211	0.172	0.149

Table 11.2 Probabilities of imaginary roots.

n	5	10	15	20
M_n	454	3332	10,878	25,336
Q_n	0.375	0.378	0.377	0.377

Some values M_n and Q_n can be found in Table 11.2.

If a, b, and c were independent random variables with the continuous distribution on the interval $[-n, n]$ (in this case n need not be an integer), we would get the probability Q_n^* of the event (11.2) in the following way. The event (11.2) occurs if either $a > 0, c > 0$, or $a < 0, c < 0$. Define

$$I_1 = \iiint_{D_1} \mathrm{d}a\, \mathrm{d}b\, \mathrm{d}c, \qquad I_1 = \iiint_{D_2} \mathrm{d}a\, \mathrm{d}b\, \mathrm{d}c$$

where

$$\begin{aligned} D_1 &= \{-n \le b \le n,\ \ 0 \le a \le n,\ \ 0 \le c \le n,\ \ b^2 < 4ac\}, \\ D_2 &= \{-n \le b \le n,\ -n \le a < 0,\ -n \le c < 0,\ b^2 < 4ac\}. \end{aligned}$$

Calculation gives

$$I_1 = 2\int_0^n \left[\int_{\frac{b^2}{4n}}^n \left(\int_{\frac{b^2}{4c}}^n \mathrm{d}a\right) \mathrm{d}c\right] \mathrm{d}b = \frac{31}{18}n^3 - \frac{1}{6}n^3 \ln 4, \quad I_2 = I_1.$$

Finally, we have

$$Q_n^* = \frac{I_1 + I_2}{(2n)^3} = \frac{31 - \ln 64}{72} = 0.373.$$

It may be somewhat unexpected that Q_n^* does not depend on n. But this point is quickly elucidated if we realize that equations $ax^2+bx+c = 0$ and $\lambda ax^2+\lambda bx+\lambda c = 0$ have the same roots for arbitrary $\lambda > 0$. Random variables λa, λb, and λc are independent and each of them has the continuous rectangular distribution on the interval $[-\lambda n, \lambda n]$.

Some other problems can be solved similarly as the problem of imaginary roots. For example

- Find the probability that at least one root of the quadratic equation is positive.

- Find the probability that both roots of the quadratic equation are positive.
- Find the probability that both roots of the quadratic equation are inside the unit circle.

Let us mention that polynomials of higher order with random coefficients are also investigated in the statistical literature, see Bharucha-Reid and Sambandham (1986), Farahmand (1991), Li (1988), Mishra et al. (1982, 1983), Sambandham et al. (1983), Sankaranarayanan (1979), and Shenker (1981).

11.2 SUM AND PRODUCT OF RANDOM NUMBERS

Let $a > 0$ be a given number. Consider random variables X and Y, which are independent and each of which has the continuous rectangular distribution on the interval $[-a, a]$. Find the probability that their *product* will exceed their *sum*. We are looking for the probability $P(a) = \mathsf{P}(XY > X + Y)$. Inequality is equivalent to $Y(X-1) > X$. The event we are considering can be written in the form

$$\begin{aligned} Y &< 1 + \frac{1}{X-1} \quad \text{for} \quad X < 1, \\ Y &> 1 + \frac{1}{X-1} \quad \text{for} \quad X > 1. \end{aligned}$$

Probability $P(a)$ is ratio of the shaded area in Fig. 11.3 to the area $2a \times 2a = 4a^2$ of the whole square. The shaded area in the first quadrant is taken into account only in the case when $a/(a-1) < a$. This occurs if $a > 2$. For $0 < a \leq 2$, we have

$$P(a) = P_1(a) = \frac{1}{4a^2} \int_{-a}^{a/(a+1)} \left(a + 1 + \frac{1}{x-1} \right) dx,$$

which gives

$$P_1(a) = \frac{a^2 + 2a - 2\ln(a+1)}{4a^2}.$$

For $a \geq 2$, the shaded area in the first quadrant is

$$Q(a) = \left(a - \frac{a}{a-1} \right) - \int_{a/(a-1)}^{a} \left(1 + \frac{1}{x-1} \right) dx = a^2 - 2a - 2\ln(a-1).$$

Thus in the case $a \geq 2$ we get

$$P(a) = P_2(a) = P_1(a) + \frac{Q(a)}{4a^2} = \frac{2a^2 - 2\ln(a^2-1)}{4a^2}.$$

The function $P(a)$ is plotted in graph form in Fig. 11.4. We can easily calculate that

$$\lim_{a \to 0+} P(a) = \frac{1}{2}, \qquad \lim_{a \to \infty} P(a) = \frac{1}{2}.$$

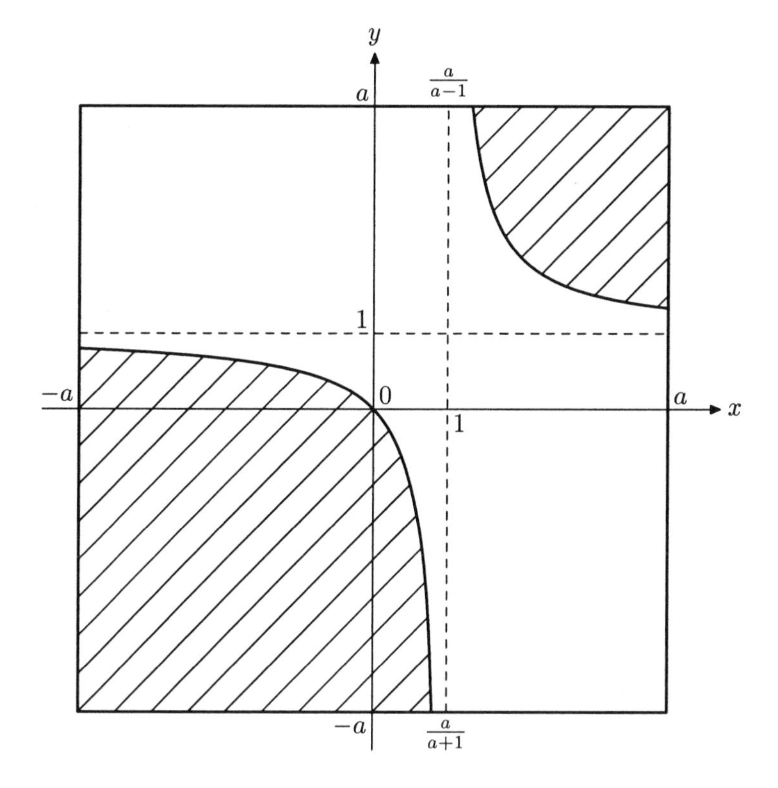

Fig. 11.3 Calculation of probability.

We analyze function $P(a)$. We roughly follow the procedure introduced in *Am. Math. Monthly* (**71**, 1961, pp. 438–439). Function $P_1(a)$ is defined for $a > 0$; function $P_2(a)$, for $a > 1$. Since $P(a) = P_1(a)$ on $(0, 2]$ and $P(a) = P_2(a)$ on $(2, \infty)$, function $P(a)$ is continuous. Define

$$f(a) = 2\ln(a+1) - a - 1 + \frac{1}{a+1}, \qquad a > -1.$$

Then $P_1'(a) = f(a)/(2a^3)$ on $(0, \infty)$ and $f'(a) = -a^2/(a+1)^2$ for $a > -1$. Since $f(0) = 0$, we have $f'(a) < 0$ on $(0, \infty)$, and thus function $P_1(a)$ is decreasing on $(0, \infty)$. Further define

$$g(a) = \ln(a^2 - 1) - \frac{a^2}{a^2 - 1}, \qquad a > 1.$$

Then $P_2'(a) = g(a)/a^3$. Since $g'(a) = 2a^3/(a^2-1)^2$ on $(1, \infty)$, function $g(a)$ is increasing on $(1, \infty)$ and obviously $g(a) \to -\infty$ as $a \to 1^+$ and $g(a) \to \infty$ as $a \to \infty$. This implies that there exists a unique $a_0 \in (1, \infty)$ such that $g(a) < 0$ on $(1, a_0)$ and $g(a) > 0$ on (a_0, ∞). Hence $P_2(a)$ is decreasing on $(1, a_0)$ and increasing on (a_0, ∞). Since $g(2) = \ln 3 - \frac{4}{3} < 0$, we have $a_0 > 2$. Then $P(a)$ is

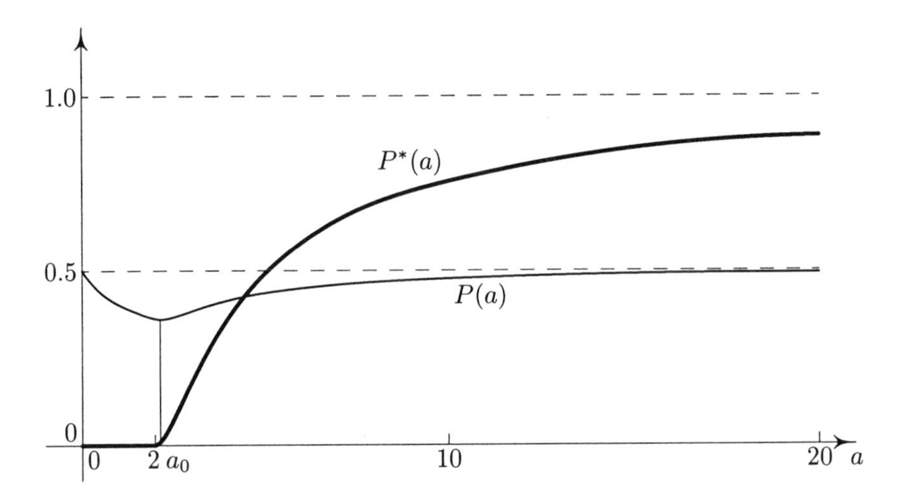

Fig. 11.4 Probabilities $P^*(a)$ and $P(a)$.

decreasing on $(0, a_0)$ and increasing on (a_0, ∞), so that it has at point a_0 absolute minimum. The equation $g(a) = 0$ can be rewritten in the form

$$\ln(a^2 - 1) = \frac{a^2}{a^2 - 1}.$$

Define

$$x = \frac{a^2}{a^2 - 1},$$

so that

$$x = \frac{a^2}{a^2 - 1}, \qquad a = \sqrt{\frac{x}{x - 1}}, \qquad a^2 - 1 = \frac{1}{x - 1}.$$

After an elementary arrangement we obtain the equation

$$x = 1 + e^{-x}.$$

If we denote its solution by x_0, then $a_0 = \sqrt{x_0/(x_0 - 1)}$. Numerically, we get $x_0 = 1.278, a_0 = 2.143, P_2(a_0) = 0.361$.

Modify the original problem to the case when X and Y are independent random variables with the continuous rectangular distribution on $[0, a]$. Again, find the probability

$$P^*(a) = \mathsf{P}(XY > X + Y).$$

In this case $P^*(a)$ is the ratio of the shaded area in the first quadrant in Fig. 11.3 to the area a^2 of the square in the first quadrant. Thus we have

$$P^*(a) = \begin{cases} 0 & \text{for} \quad 0 < a \leq 2, \\ \dfrac{Q(a)}{a^2} = \dfrac{a^2 - 2a - 2\ln(a - 1)}{a^2} & \text{for} \quad a > 2. \end{cases}$$

We can see that $P^*(a) \to 1$ as $a \to \infty$. If $a > 2$, then we compute

$$\frac{dP^*(a)}{da} = \frac{4\ln(a-1) + \frac{2a(a-2)}{a-1}}{a^3} > 0,$$

so that $P^*(a)$ is increasing on the interval $(2, \infty)$. This is also depicted graph form in Fig. 11.4.

11.3 SOCKS AND NUMBER THEORY

Red and black *socks* are in a draw of the chest of drawers. It is dark and socks are mixed. We randomly select two socks. It is known that the probability that the two socks is a red pair is exactly $\frac{1}{2}$. Find the number of red and black socks in the drawer.

This problem is introduced in the collection Mosteller (1965), where there is an additional demand that the number of all socks in the drawer is minimal. Another variant of the problem is that the number of black socks must be even.

Let r and b be numbers of red and black socks in the drawer, respectively. From the condition that probability of selecting two red socks is $\frac{1}{2}$, we obtain the fundamental equation

$$\frac{r}{r+b}\,\frac{r-1}{r+b-1} = \frac{1}{2}. \tag{11.3}$$

It is clear that $r > b$. In the case $r \le b$ we would have

$$r = \frac{r}{2} + \frac{r}{2} \le \frac{r}{2} + \frac{b}{2} = \frac{1}{2}(r+b).$$

This would imply that $r/(r+b) \le \frac{1}{2}$ and analogously we would get the inequality $(r-1)/(r-1+b) < \frac{1}{2}$. Then the left-hand side of the formula (11.3) would be smaller than $\frac{1}{4}$.

From (11.3) we easily obtain an equivalent equation:

$$(r-b)(r-b-1) = 2b^2. \tag{11.4}$$

Define

$$r - b - 1 = x, \qquad b = y. \tag{11.5}$$

Then (11.4) can be written in the form

$$x^2 + x - 2y^2 = 0. \tag{11.6}$$

As soon as we know integer-valued solutions of this equation, using (11.5), we obtain

$$r = x + y + 1, \qquad b = y. \tag{11.7}$$

It is evident that

$$x_1 = y_1 = 1 \tag{11.8}$$

Table 11.3 Number of socks r and b.

x	y	r	b	$r+b$
1	1	3	1	4
8	6	15	6	21
49	35	85	35	120
288	204	493	204	697
1,681	1,189	2,871	1,189	4,060
9,800	6,930	16,731	6,930	23,661

is a solution of the equation (11.6). Further, we easily verify the following assertion. If x_n, y_n $(n \geq 1)$ is a solution of the equation (11.6), then

$$x_{n+1} = 3x_n + 4y_n + 1, \qquad y_{n+1} = 2x_n + 3y_n + 1 \qquad (n \geq 1) \tag{11.9}$$

is also a solution of this equation. Indeed

$$\begin{aligned} x_{n+1}^2 + x_{n+1} - 2y_{n+1}^2 &= (3x_n + 4y_n + 1)^2 + 3x_n + 4y_n + 1 \\ &\quad -2(2x_n + 3y_n + 1)^2 \\ &= x_n^2 + x_n - 2y_n^2 = 0. \end{aligned}$$

It can be shown that all positive integer-valued solutions of equation (11.6) are given by (11.8) and (11.9). The proof is longer, but not difficult and can be found in the book by Sierpiński (1956, Sect. 5) (there is also a Russian translation from 1961).

All solutions of equation (11.6) can be calculated successively from (11.8) and (11.9). Inserting them into (11.6), we have an answer to the question. Some solutions are introduced in Table 11.3. The drawer would not hold a larger number of socks.

Now we can also answer questions asked by Mosteller. The smallest number of socks that enables us to solve the problem is 4. This may be a nontypical case, since it is rare to tidy up three red socks and one black sock. The smallest number of socks when the number of the black ones is even is 21. Then there are 15 red and 6 black socks in the drawer.

A few concluding remarks to the problem. Equation (11.6) can also be written in the form $\binom{x+1}{2} = y^2$. The problem is to determine when the combinatorial number $\binom{x+1}{2}$ is the square of an integer y. If we were interested only in numbers y (and not in numbers x), we could write the second part of the formula (11.9) in the form

$$y_{n+2} = 2x_{n+1} + 3y_{n+1} + 1. \tag{11.10}$$

Thus we obtain the following equations successively from (11.9):

$$x_n = \frac{1}{2}(y_{n+1} - 3y_n - 1), \qquad x_{n+1} = \frac{3}{2}(y_{n+1} - 3y_n - 1) + 4y_n + 1.$$

Inserting the last expression into (11.10), we get

$$y_{n+2} = 6y_{n+1} - y_n, \qquad n \geq 1. \tag{11.11}$$

We can continue in the following way. With respect to (11.8), we know from (11.9) that $x_2 = 8$, $y_2 = 6$. Then we know that $y_1 = 1$, $y_2 = 6$, and we calculate further values y_n from (11.11). Incidentally, (11.11) is a homogeneous linear difference equation of the second order. The corresponding characteristic equation $\lambda^2 - 6\lambda + 1 = 0$ has roots $\lambda_{12} = 3 \pm \sqrt{8}$. Thus the general solution of the equation (11.11) is

$$y_n = A(3+\sqrt{8})^n + B(3-\sqrt{8})^n,$$

where A and B are some constants. From the initial conditions $y_1 = 1$, $y_2 = 6$ (or even faster from the conditions $y_0 = 0$, $y_1 = 1$), we get $A = -B = 1/(2\sqrt{8})$, and thus

$$y_n = \frac{1}{2\sqrt{8}}\left[(3+\sqrt{8})^n - (3-\sqrt{8})^n\right].$$

There exist an infinite number of binomial coefficients of the type $\binom{x+1}{2}$ that are squares of positive integers. On the other hand, it was proved that with the exception of $x = 1$, no binomial coefficient $\binom{x+1}{2}$ is the fourth power of an integer; the equation $x^2 + x - 2y^4 = 0$ has no solution in positive integers larger than 1. However, it has solutions in rational positive numbers, for example $x = \frac{32}{49}$, $y = \frac{6}{7}$.

Another approach to the solution of the original problem can be based on the following idea. We know from (11.6) that

$$x = \frac{-1+\sqrt{1+8y^2}}{2}.$$

However, x is a positive integer if and only if $1+8y^2$ is a square, such as $1+8y^2 = z^2$. But $z^2 - 8y^2 = 1$ is *Pell's equation* (see Sierpiňski 1956, Sect. 7; Valfiš 1952). General number theory leads to the following result. If a_n and b_n are positive integers defined by

$$a_n + b_n\sqrt{8} = (3+\sqrt{8})^n, \qquad n = 1, 2, \ldots,$$

and if we define $c_n = (a_n + 1)/2$, then

$$\binom{c_n}{2} = b_n^2$$

are all solutions of the equation $\binom{c}{2} = b^2$ in positive integers. Frankly speaking, only the general method of solution of Pell's equation answers the question of how solution (11.9) was derived.

11.4 TSHEBYSHEV PROBLEM

Formulation of the *Tshebyshev problem* is simple. Calculate the probability that two randomly chosen positive integers are relatively prime. However, the set of positive integers has infinitely many elements, and it is not completely clear what is understood by the term *random choice*. Thus we shall state it more explicitly. Let M be a given positive integer. Let X and Y be independent random variables with discrete rectangular distribution on the set $\{1, 2, \ldots, M\}$. Denote by p_M the probability that the numbers X and Y are relatively prime. Find $p = \lim_{M\to\infty} p_M$.

Let $2, 3, 5, 7, \ldots, p_k$ be first k *prime numbers*. Each number M can be written in the form $M = N+n$, where N is a multiple of the product $s_k = 2\times 3\times 5\times 7\times\cdots\times p_k$ and $0 \le n < s_k$. The probability that both numbers X and Y are less than or equal to N is $(N/M)^2$. Then the probability that at least one of the numbers X and Y is larger than N is

$$1 - \frac{N^2}{M^2} = \frac{(M-N)(M+N)}{M^2} \le \frac{s_k 2N}{M^2} = 2\frac{s_k}{M} \to 0$$

for every fixed k as $M \to \infty$.

Further, we consider only numbers X and Y that are less than or equal to N. We have N^2 of pairs (X, Y). Only $N^2/4$ of them are such that X and Y are even. There exist $N^2/9$ pairs such that X and Y are divisible by 3. However, $N^2/36$ of them are pairs such that X and Y are even. This implies that there exist

$$\frac{N^2}{9} - \frac{N^2}{36} = \frac{N^2}{9}\left(1 - \frac{1}{4}\right)$$

pairs (X, Y) such that X and Y are divisible by 3 but not by 2. The number of pairs (X, Y), such that their common divisor is neither 3 nor 2, is

$$N^2 - \frac{N^2}{4} - \frac{N^2}{9}\left(1 - \frac{1}{4}\right) = N^2\left(1 - \frac{1}{4}\right)\left(1 - \frac{1}{9}\right).$$

Other prime divisors can be considered in an analogous fashion. We obtain

$$p_N = \left(1 - \frac{1}{4}\right)\left(1 - \frac{1}{9}\right)\cdots\left(1 - \frac{1}{p_k^2}\right).$$

This number does not depend on N. Since the difference between p_N and p_M tends to zero as $M \to \infty$, we have for every k that

$$\lim_{M\to\infty} p_M = \left(1 - \frac{1}{4}\right)\left(1 - \frac{1}{9}\right)\cdots\left(1 - \frac{1}{p_k^2}\right).$$

If we also let $k \to \infty$, then we obtain

$$p = \prod_{k=1}^{\infty}\left(1 - \frac{1}{p_k^2}\right).$$

In view of

$$\frac{1}{1 - \dfrac{1}{p_k^2}} = 1 + \frac{1}{p_k^2} + \frac{1}{p_k^4} + \frac{1}{p_k^6} + \cdots,$$

after multiplying and rearranging terms, we get

$$\frac{1}{p} = \sum_{k=1}^{\infty} \frac{1}{k^2} = \frac{\pi^2}{6}.$$

Thus the probability that two randomly chosen positive integers are relatively prime is

$$p = \frac{6}{\pi^2} = 0.608.$$

A detailed derivation can be found in the book by Jaglom and Jaglom (1954, pp. 307–315; solution of Problems 89 and 90). In the same book it is proved that four randomly chosen positive integers have a nontrivial common divisor with probability

$$1 - \frac{90}{\pi^4} = 0.075.$$

11.5 RANDOM TRIANGLE

Three points, A, B, C, are randomly chosen in a plane. Find probability P_a that the *triangle ABC* is acute. This problem was proposed and solved in Woolhouse (1886). The solution published there was

$$P_a = \frac{4}{\pi^2} - \frac{1}{8} \doteq 0.280.$$

It was shown later (Eisenberg and Sullivan 1996) that this is a correct solution of another problem, namely, when the three vertices are independent uniformly distributed points in the unit disk. The original formulation is not quite correct. If we say "choose three points randomly in a given set," we usually mean that the three points are chosen independently from the rectangular distribution on the set. The plane does not have a finite Lebesgue measure, and so we cannot consider any rectangular distribution on the whole plane.

Charles Dodgson (1832–1898), who was known under the pen name Lewis Carroll, was the author of the books *Alice's Adventures in Wonderland* and *Through the Looking Glass*. In his work Carroll (1893) tried to solve the problem in the following way (see Portnoy 1994). Let AB denote the longest side of the triangle. Should it be the longest one, then the vertex C must lie in the intersection Q of disks K_1 and K_2, which have the same radius $r = AB$ and centers at A and B (see Fig. 11.5). Let S be the midpoint of the segment AB. The angle at the vertex C is obtuse if and only if C lies inside the disk K with the center S and diameter AB (remember the *Thalet theorem*). Then the probability $P_o = 1 - P_a$, that the triangle is obtuse, should be given by geometric probability as the ratio of the area of the disk K to the area of the region Q. The area of the disk K is $\pi r^2/4$, the area of Q is $(2\pi/3 - \sqrt{3}/2)r^2$, and thus their ratio is

$$P_o = \frac{3\pi}{8\pi - 6\sqrt{3}} \doteq 0.639.$$

This yields $P_a = 1 - P_o \doteq 0.361$. The solution is dubious, since it does not follow from the formulation of the problem that point C has the rectangular distribution on the region Q.

We show that similar arguments lead to completely different results (see Portnoy 1994 and Guy 1993).

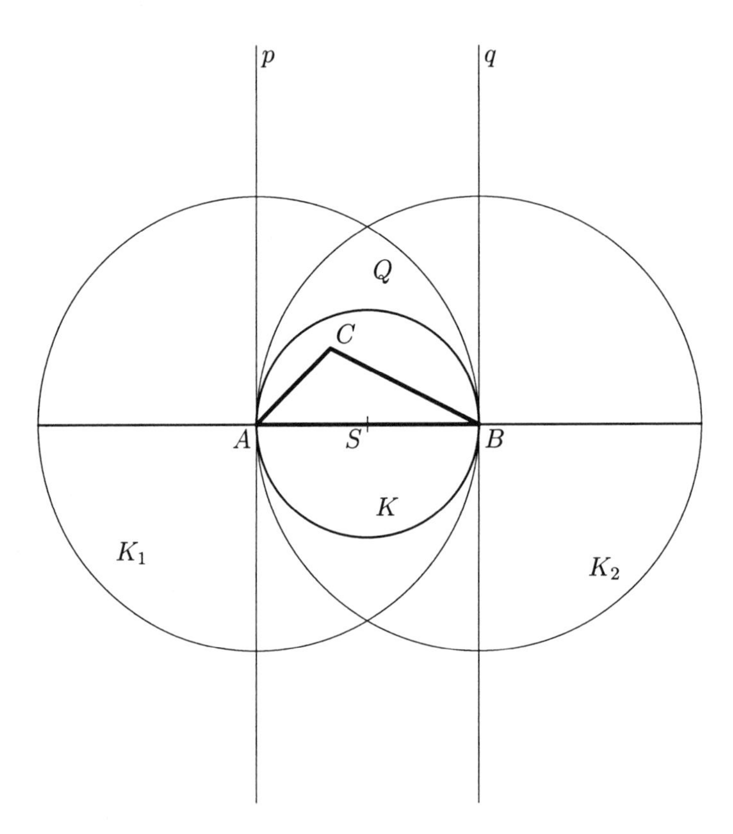

Fig. 11.5 Carroll's method.

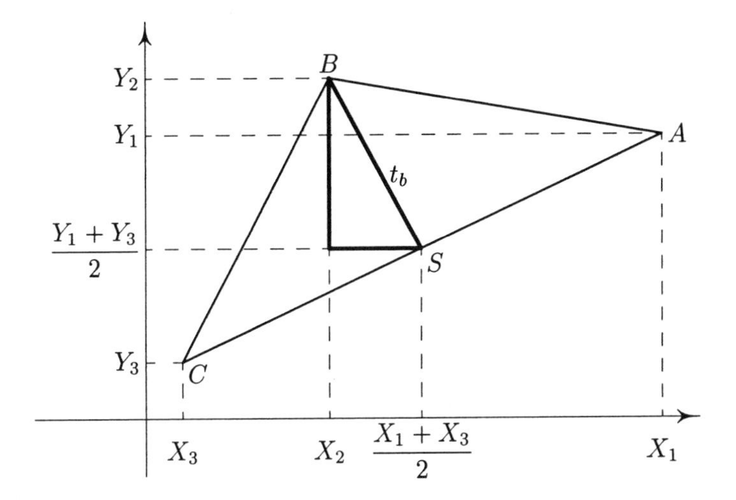

Fig. 11.6 Gaussian triangle.

If segment AB were the secondlongest side of the triangle, then point C would have to lie in union $K_1 \cup K_2$ but outside the region Q (i.e., in the symmetric difference $K_1 \div K_2$.). The triangle is obtuse if C lies outside the strip bounded by lines p and q. The area of the intersection of the outside region of the strip and $K_1 \div K_2$ is πr^2, and the area of $K_1 \div K_2$ is $(2\pi/3 + \sqrt{3})r^2$. Geometric probability defines P_o as the ratio of both areas, and we obtain

$$P_o = \frac{3\pi}{2\pi + 3\sqrt{3}} \doteq 0.821.$$

In this case we used an unjustified assumption that C has the rectangular distribution on the region $K_1 \div K_2$.

If AB were the shortest side of the triangle, point C would have to lie outside $K_1 \cup K_2$. The triangle ABC is obtuse if and only if C lies in the same time outside the strip defined by the lines p and q. Both the abovementioned regions have infinite area, and geometric probability is not defined.

Guy (1993) introduces some other interpretations of the original problem. We have an analogy with the Bertrand paradox, although the incorrect formulation has a character deeper than that of nonuniqueness by Bertrand.

Consider a modern formulation of the problem. Let

$$A = (X_1, Y_1), \quad B = (X_2, Y_2), \quad C = (X_3, Y_3).$$

Assume that all variables $X_1, X_2, X_3, Y_1, Y_2, Y_3$ are independent and each of them has the normal distribution $\mathsf{N}(0, 1)$. Triangle ABC is called the *Gaussian triangle.* Maximally, one of the vertices of the triangle can be obtuse. Moreover, the joint distribution of the coordinates of vertices is invariant under permutation of the variables

$X_1, X_2, X_3, Y_1, Y_2, Y_3$. This implies that

$$\begin{aligned} P_o &= \mathsf{P}(\Delta ABC \text{ is obtuse}) \\ &= \mathsf{P}(\angle ABC > 90^\circ \text{ or } \angle CAB > 90^\circ \text{ or } \angle BCA > 90^\circ) \\ &= 3\mathsf{P}(\angle ABC > 90^\circ). \end{aligned}$$

But $\angle ABC$ is obtuse if and only if the median t_b from B to the midpoint S of the side AC is smaller than half of the segment AC. The proof is analogous to Carroll's procedure. We can see in Fig. 11.6 that

$$\begin{aligned} t_b^2 &= \left(X_2 - \frac{X_1 + X_3}{2}\right)^2 + \left(Y_2 - \frac{Y_1 + Y_3}{2}\right)^2, \\ |SC|^2 &= \left(\frac{X_1 - X_3}{2}\right)^2 + \left(\frac{Y_1 - Y_3}{2}\right)^2. \end{aligned}$$

We introduce variables

$$\begin{aligned} U_1 &= X_2 - \frac{X_1 + X_3}{2}, \quad U_2 = Y_2 - \frac{Y_1 + Y_3}{2}, \\ V_1 &= \frac{X_1 - X_3}{2}, \quad V_2 = \frac{Y_1 - Y_3}{2} \end{aligned}$$

and vectors

$$\boldsymbol{\xi} = (X_1, X_2, X_3, Y_1, Y_2, Y_3)', \qquad \boldsymbol{\eta} = (U_1, U_2, V_1, V_2)'.$$

We easily see that $\boldsymbol{\eta} = \boldsymbol{H}\boldsymbol{\xi}$ where

$$\boldsymbol{H} = \begin{pmatrix} -\frac{1}{2} & 1 & -\frac{1}{2} & 0 & 0 & 0 \\ 0 & 0 & 0 & -\frac{1}{2} & 1 & -\frac{1}{2} \\ \frac{1}{2} & 0 & -\frac{1}{2} & 0 & 0 & 0 \\ 0 & 0 & 0 & \frac{1}{2} & 0 & -\frac{1}{2} \end{pmatrix}.$$

Since $\boldsymbol{\xi} \sim N_6(\mathbf{0}, \boldsymbol{I})$, we have $\boldsymbol{\eta} \sim N_4(\mathbf{0}, \boldsymbol{H}\boldsymbol{H}')$ (see Anděl 1978, Theorem V.4). Here $\boldsymbol{H}\boldsymbol{H}' = \mathsf{Diag}\{\frac{3}{2}, \frac{3}{2}, \frac{1}{2}, \frac{1}{2}\}$. Thus the variables U_1, U_2, V_1, V_2 are independent, normally distributed with vanishing expectations, and their variances are $\frac{3}{2}, \frac{3}{2}, \frac{1}{2}, \frac{1}{2}$, respectively. Thus

$$\begin{aligned} P_o &= 3\mathsf{P}(U_1^2 + U_2^2 < V_1^2 + V_2^2) \\ &= 3\mathsf{P}\left\{\frac{3}{2}\left[\left(U_1\sqrt{\frac{2}{3}}\right)^2 + \left(U_2\sqrt{\frac{2}{3}}\right)^2\right] < \frac{1}{2}[(V_1\sqrt{2})^2 + (V_2\sqrt{2})^2]\right\}. \end{aligned}$$

Define

$$W = \left(U_1\sqrt{\frac{2}{3}}\right)^2 + \left(U_2\sqrt{\frac{2}{3}}\right)^2, \qquad Z = (V_1\sqrt{2})^2 + (V_2\sqrt{2})^2.$$

Variables W and Z are independent, and each of them has distribution χ_2^2 (see Anděl 1978, Theorem V.10). The ratio W/Z has $\mathsf{F}_{2,2}$ distribution (see Anděl 1978, Theorem V.22) with density $f_{2,2}(z)$ and distribution function $F_{2,2}(z)$, which vanish for $z \leq 0$ and are defined by the formulas

$$f_{2,2}(z) = (1+z)^{-2}, \qquad F_{2,2}(z) = \frac{z}{1+z}$$

if $z > 0$. Thus we obtain

$$P_o = 3\mathsf{P}\left(\frac{3}{2}W < \frac{1}{2}Z\right) = 3\mathsf{P}\left(\frac{W}{Z} < \frac{1}{3}\right) = 3F_{2,2}\left(\frac{1}{3}\right) = \frac{3}{4}. \tag{11.12}$$

We add a few remarks. It can be proved (see Portnoy 1994) that the result $P_o = \frac{3}{4}$ is valid not only for Gaussian triangles but also under a more general condition when $X_1, X_2, X_3, Y_1, Y_2, Y_3$ have *spherically symmetric distribution* in $\mathbb{R}^6$.

The results were generalized to random triangles in $\mathbb{R}^n$ (Hall 1982, Eisenberg and Sullivan 1996). For example, the probability that a Gaussian triangle in $\mathbb{R}^n$ is obtuse equals

$$P_o^{\{n\}} = \frac{3\Gamma(n)}{\left[\Gamma\left(\frac{n}{2}\right)\right]^2} \int_0^{1/3} \frac{x^{(n-2)/2}\,\mathrm{d}x}{(1+x)^n}. \tag{11.13}$$

If $n = 2k$, where k is a positive integer, this formula can be written in the form

$$P_o^{\{n\}} = 3\sum_{j=k}^{2k-1} \binom{2k-1}{j} \left(\frac{1}{4}\right)^j \left(\frac{3}{4}\right)^{2k-1-j}. \tag{11.14}$$

Of course, formula (11.12) is a special case of (11.13) and (11.14).

11.6 LATTICE-POINT TRIANGLES

Consider *lattice points* M with integer-valued coordinates in a plane. It is known that the area S of a triangle with vertices in M is always half of an integer (in some cases it is half of an even integer, in other cases half of an odd integer). It follows from Pick's[1] formula that the area of a polygon with vertices that have integer coordinates is the number of interior lattice points, plus half the number of boundary lattice points, minus 1.

Assume that three lattice points from a rectangle are chosen at random. Find the probability that the area of the triangle they determine is an integer when both sides

[1]Georg Pick was born on 10th August 1859 in Vienna. He graduated from the University of Vienna in 1880. He was appointed Associated Professor at Prague German University in 1882, where he was Extraordinary Professor from 1888 to 1892, and was appointed Full Professor in 1892. He was Dean of the Faculty of Arts in the academic year 1900/01. On July 13, 1942, at age 83 years, he was deported from Prague to the concentration camp Theresienstadt. He died in Theresienstadt on July 26, 1942. See Pinl and Dick (1974), Ludvíková (1997), and Netuka (1999). His formula, mentioned here, was published in Pick (1899).

of the rectangle grow to infinity. This problem was published in *Math. Magazine* (**57**, 1984, p. 238), with its solution in the same journal (**58**, 1985, p. 240).

If the vertices of the triangle are (a_1, b_1), (a_2, b_2), (a_3, b_3), then its area is equal to the absolute value of

$$\frac{1}{2}\begin{vmatrix} a_1 & b_1 & 1 \\ a_2 & b_2 & 1 \\ a_3 & b_3 & 1 \end{vmatrix} = \frac{1}{2}[(a_2 - a_1)(b_3 - b_1) - (a_3 - a_1)(b_2 - b_1)]$$

(see Rektorys et al. 1995, Vol. I, p. 95). From here we can also easily see that the area of a lattice triangle is half of an integer. Assume that the vertices of the triangle are randomly chosen in lattice points of the rectangle $|a_i| \leq A$, $|b_i| \leq B$. Define

$$a_2' = a_2 - a_1, \quad a_3' = a_3 - a_1, \quad b_2' = b_2 - b_1, \quad b_3' = b_3 - b_1.$$

Then $2S = |a_2'b_3' - a_3'b_2'|$. We can see that

$$\begin{aligned} \mathsf{P}(2S \text{ is even}) &= \mathsf{P}(a_2'b_3' \text{ is even and } a_3'b_2' \text{ is even}) \\ &\quad + \mathsf{P}(a_2'b_3' \text{ is odd and } a_3'b_2' \text{ is odd}). \end{aligned}$$

This expression tends to

$$\frac{3}{4} \times \frac{3}{4} + \frac{1}{4} \times \frac{1}{4} = \frac{5}{8}$$

as $A, B \to \infty$.

12

Matrix games

12.1 LINEAR PROGRAMMING

As a motivation, we introduce a *nutrition problem*, which is also called a *"diet problem"*. Animals need n different nutrients. The nutrients are contained in m foods. It is known that

1. In the ith food, a_{ij} units of jth nutrients ($i = 1, \ldots, m$; $j = 1, \ldots, n$) are contained.

2. In a given time interval, the animals must have at least c_j units (e.g., kilograms) of the jth nutrient ($j = 1, \ldots, n$).

3. The cost per unit of the ith food is b_i ($i = 1, \ldots, m$).

We must decide how many units of individual foods must be bought to fulfill requirement 2 in such a way that the total cost of all foods is minimal. Let y_i be the number of purchased units of the ith food ($i = 1, \ldots, m$). The cost of this purchase is

$$f(y) = \sum_{i=1}^{m} b_i y_i. \tag{12.1}$$

To fulfill condition 2, we must have

$$\sum_{i=1}^{m} y_i a_{ij} \geq c_j, \qquad j = 1, \ldots, n. \tag{12.2}$$

The formulation of the problem implicitly contains additional important constraints. The number of purchased units of individual foods cannot be negative. Hence we must add the constraints

$$y_i \geq 0, \qquad i = 1, \ldots, m. \tag{12.3}$$

Function f defined in (12.1) is called the *objective function*. A matrix notation simplifies formulas introduced above. Define

$$\boldsymbol{A} = (a_{ij})_{i=1}^{m}{}_{j=1}^{n}, \qquad \boldsymbol{b} = \begin{pmatrix} b_1 \\ \vdots \\ b_m \end{pmatrix}, \qquad \boldsymbol{c} = \begin{pmatrix} c_1 \\ \vdots \\ c_n \end{pmatrix}, \qquad \boldsymbol{y} = \begin{pmatrix} y_1 \\ \vdots \\ y_m \end{pmatrix}.$$

If $\boldsymbol{u} = (u_1, \ldots, u_k)'$ and $\boldsymbol{v} = (v_1, \ldots, v_k)'$ are two vectors with the same number of components, then we write $\boldsymbol{u} \geq \boldsymbol{v}$ in the case that $u_i \geq v_i$ for each $i = 1, \ldots, k$. Similarly we write $\boldsymbol{u} > \boldsymbol{v}$, if $u_i > v_i$ for each $i = 1, \ldots, k$. The nutrition problem can be formulated in the following way:

$$\text{Minimize } \boldsymbol{b}'\boldsymbol{y} \text{ under constraints } \boldsymbol{y}'\boldsymbol{A} \geq \boldsymbol{c}', \ \boldsymbol{y} \geq \boldsymbol{0}. \tag{12.4}$$

The nutrition problem was solved in the United States in 1944. The optimal result was perhaps suitable for animals, but people would surely refuse such a menu. The resulting combination of different foods was reported to be untasteful (see Karlin 1959).

The nutrition problem was based on minimization of cost. Other problems call for maximization of production or sale under some technological restrictions. If we define $\boldsymbol{x} = (x_1, \ldots, x_n)'$, then the mathematical formulation of such a problem will be

$$\text{Maximize } \boldsymbol{c}'\boldsymbol{x} \text{ under constraints } \boldsymbol{A}\boldsymbol{x} \leq \boldsymbol{b}, \ \boldsymbol{x} \geq \boldsymbol{0}. \tag{12.5}$$

Mathematically, both kinds of problems are equivalent. Formulation (12.4) can be equivalently written in the form

$$\text{Maximize } (-\boldsymbol{b}')\boldsymbol{y} \text{ under constraints } (-\boldsymbol{A})'\boldsymbol{y} \leq -\boldsymbol{c}, \ \boldsymbol{y} \geq \boldsymbol{0}.$$

In a general case it is not necessary that all components of the vector $\boldsymbol{x}$ or $\boldsymbol{y}$ be nonnegative. Sometimes some inequalities from the system $\boldsymbol{A}\boldsymbol{x} \leq \boldsymbol{b}$ (or $\boldsymbol{y}'\boldsymbol{A} \geq \boldsymbol{c}'$) must also be fulfilled as equalities.

Problems (12.4) and (12.5) and their generalizations are called *linear programming problems*. In statistical papers and books (e.g., Dupačová 1982, Karlin 1959) conditions are introduced under which the problem has a solution. Algorithms are described to enable the reader to find a solution. The most frequently used algorithm is the *simplex method*.

It was proved that a solution must lie in an extreme point of the set defined by constraints introduced in formulas (12.4) and (12.5). From a geometric point of view, the set is simplex and so it suffices to analyze values of the objective function $\boldsymbol{b}'\boldsymbol{y}$ or

$c'x$ only in vertices of the simplex. However, this is still very difficult, because real problems lead to simplexes with so many vertices that even modern computers are not able to take into account all of them in a reasonable time. Numerical algorithms start at a vertex but continue at another one, which gives a more advantageous value of the objective function. In this book we consider linear programming only as a method that enables us to solve problems such as (12.4) and (12.5) on a computer. We use it as an instrument for solving some probabilistic and statistical problems.

12.2 PURE STRATEGIES

We consider only games with two players, say, A and B. Cases involving three or more players are more complicated, since the players can form coalitions and work as teams. Moreover, we restrict our presentation to *zero-sum games*, in which the amount that one of the players wins equals the amount that the other player loses. Each player has several possibilities, called *strategies*, and chooses one of them. For simplicity, we introduce only such games where the number of strategies is finite. Assume that player A can choose one of strategies $A_1, \ldots, A_m$ and player B one of strategies $B_1, \ldots, B_n$. Let a_{ij} be the amount paid to A, in the case that A selects strategy A_i and B selects strategy B_j. If $a_{ij} < 0$, then A pays amount $|a_{ij}|$. Strategies A_i and B_j are called *"pure" strategies*. The matrix

$$\boldsymbol{A} = \begin{pmatrix} a_{11} & \cdots & a_{1n} \\ \ldots & \ldots & \ldots \\ a_{m1} & \cdots & a_{mn} \end{pmatrix} \tag{12.6}$$

is called *"payoff" matrix*. In some cases $\boldsymbol{A}$ has a simple structure and it is easy to find optimal strategies. We demonstrate this in an example. Consider the payoff matrix

$$\boldsymbol{A} = \begin{pmatrix} 7 & 2 & 5 & 1 \\ 2 & 2 & 3 & 4 \\ 5 & 3 & 4 & 4 \\ 3 & 2 & 1 & 6 \end{pmatrix}. \tag{12.7}$$

If player A selects the first strategy and B the third one, then A obtains a prize of 5 units. For the moment we ignore the fact that the play is unfair for B, who must only pay and pay. Everybody can be in such a situation — for example, everybody must pay taxes. But in the frame of given laws we can select such a strategy that minimizes the tax to be paid.

We return to the problem of how to find the optimal strategy for each player. By selecting strategy A_1, A can win even 7 (if B selects strategy B_1). But A can also win only 1 (if B selects B_4). Thus strategy A_1 ensures that A wins at least 1. Similarly, A_2 ensures that A wins at least 2, A_3 at least 3, and A_4 at least 1. From this point of view A_3 is the best strategy since it ensures a prize of at least 3, and this cannot ensure any other strategy.

By selecting strategy B_1 player B risks paying 7 in the worst case. By selecting B_2 B pays maximally 3; selecting B_3, maximally 5; and selecting B_4, maximally 6.

B's risk is minimal under B_2, and its value is 3. The elements $-a_{ij}$ of the matrix $-\boldsymbol{A}$ are prizes of player B. Then B_2 ensures the maximal prize for player B, although it is -3 in this case. Thus A_3 is the optimal strategy of player A and B_2 is the optimal strategy of player B. The solution was so simple since the element a_{32} is minimal in its row and maximal in its column. Such an element is called a *saddle point* of the matrix.

If a saddle point exists, then both players can derive optimal pure strategies in this way, his own optimal strategy as well as the optimal strategy of his opponent. If player A declares intention of using the optimal strategy, this information does nothing to help player B. If player A selects the optimal strategy but B does not, then B yields no advantage.

In the case of a general payoff matrix $\boldsymbol{A}$ we say that it has a saddle point if

$$\max_i \min_j a_{ij} = \min_j \max_i a_{ij}. \tag{12.8}$$

Let a_{rs} be an element such that

$$a_{rs} = \max_i \min_j a_{ij}.$$

If (12.8) holds, then a_{rs} is a saddle point and A_r, B_s are optimal strategies of the players. The number a_{rs} is called the *"pure" value of the game.* If the pure value of the game is zero, we say that the game is *fair.*

Games with a saddle point are not very interesting. However, matrix (12.7) has an important feature. By comparing strategies B_1 and B_2, player B concludes that regardless of the third and fourth columns of the payoff matrix, it would be unwise to use B_1 instead of B_2 independently of any strategy of player A. The elements of the first column are equal to or larger than the corresponding elements of the second column. Hence player B, selecting B_1, pays always at least so much as when selecting B_2. It is disadvantageous for B to use strategy B_1. We say that strategy B_2 dominates strategy B_1. Instead of the game with the payoff matrix (12.7), it suffices to analyze a *reduced matrix*

$$\boldsymbol{A}_{\mathrm{r}} = \begin{pmatrix} 2 & 5 & 1 \\ 2 & 3 & 4 \\ 3 & 4 & 4 \\ 2 & 1 & 6 \end{pmatrix}.$$

This matrix has smaller dimensions, and any further computations will be easier. Also note that strategy A_3 dominates strategy A_2. Then it is possible to reduce the payoff matrix again, this time to

$$\boldsymbol{A}_{\mathrm{rd}} = \begin{pmatrix} 2 & 5 & 1 \\ 3 & 4 & 4 \\ 2 & 1 & 6 \end{pmatrix}.$$

However, equality (12.8) may not hold in a general case. As an example, we introduce a payoff matrix

$$\boldsymbol{A} = \begin{pmatrix} 1 & -1 \\ -1 & 1 \end{pmatrix}. \tag{12.9}$$

This matrix arises in the familiar *game of matching pennies*. Player A selects either heads or tails, and player B, not knowing A's choice, also selects either heads or tails. After the choices have been made, A obtains one unit from B if they match and -1 unit if they do not.

Theorem 12.1 *Let $\boldsymbol{A}$ be a payoff matrix. Then*

$$\max_i \min_j a_{ij} \leq \min_j \max_i a_{ij}. \tag{12.10}$$

Proof. For each fixed i and r, we have

$$\min_j a_{ij} \leq a_{ir} \leq \max_s a_{sr}.$$

This inequality also holds for the i that maximizes the left-hand side. Thus

$$\max_i \min_j a_{ij} \leq \max_s a_{sr}.$$

This inequality holds for each r, including the one that minimizes the right-hand side. This concludes the proof. $\square$

12.3 MIXED STRATEGIES

Consider the matrix $\boldsymbol{A}$ introduced in (12.9). This matrix has no saddle point, since $\max_i \min_j a_{ij} = -1$, $\min_j \max_i a_{ij} = 1$. The equality (12.8) does not hold, inequality (12.10) is sharp. The pure value of the game does not exist. But anyone who has ever played the game of matching pennies knows that the use of a pure strategy is rather unfavorable. For example, if player A decided to use strategy A_1, then player B would systematically select strategy B_2 and the payoff for A would be -1. It is better to change strategies randomly.

Consider the general payoff matrix (12.6). Player A can decide to use strategy A_1 with probability u_1 and so on, and strategy A_m with probability u_m. The vector $\boldsymbol{u} = (u_1, \ldots, u_m)'$ is called a *mixed strategy* of player A. All components of the vector $\boldsymbol{u}$ must be nonnegative, and their sum must equal 1. The set of such vectors is denoted by $\mathcal{U}$. Similarly, the set of mixed strategies $\boldsymbol{v} = (v_1, \ldots, v_m)'$ is denoted by $\mathcal{V}$. In a special case one component of the mixed strategy can be equal to 1 and all others to 0. Then the mixed strategy becomes a pure strategy. On the other hand, we say that a mixed strategy is *completely mixed* if all its components are positive. In some cases this definition simplifies the computation of optimal mixed strategies.

If A selects strategy $\boldsymbol{u}$ and B selects strategy $\boldsymbol{v}$, then the expected prize of player A is

$$f(\boldsymbol{u}, \boldsymbol{v}) = \sum_{i=1}^{m} \sum_{j=1}^{n} a_{ij} u_i v_j = \boldsymbol{u}' \boldsymbol{A}\, \boldsymbol{v}.$$

Function $f(\boldsymbol{u}, \boldsymbol{v})$ is also called the *payoff function* for A. Strategy $\boldsymbol{u}$ ensures that A has an expected prize of at least $\min_{\boldsymbol{v} \in \mathcal{V}} f(\boldsymbol{u}, \boldsymbol{v})$. Player A can select from the set $\mathcal{U}$

a strategy such that it maximizes the expected prize. Then A can be sure to get the expected prize:

$$C_1 = \max_{\boldsymbol{u} \in \mathcal{U}} \min_{\boldsymbol{v} \in \mathcal{V}} f(\boldsymbol{u}, \boldsymbol{v}).$$

Similar consideration leads to the conclusion that B can be sure to have expected payment maximally:

$$C_2 = \min_{\boldsymbol{v} \in \mathcal{V}} \max_{\boldsymbol{u} \in \mathcal{U}} f(\boldsymbol{u}, \boldsymbol{v}).$$

In these formulas we should write sup instead of max and inf instead of min. It can be proved that supremum and infimum are reached and that our notation is correct.

Theorem 12.2 *We have* $C_1 \leq C_2$.

Proof. The proof is similar to that of Theorem 12.1. See Karlin (1959, Lemma 1.3.1). □

The theory of matrix games is based on the assertion that, in fact, equality holds in Theorem 12.2.

Theorem 12.3 (the fundamental theorem of matrix games) *We have* $C_1 = C_2$.

Proof. In this case the proof is more complicated. For example, it can be found in Karlin (1959, Theorem 1.4.1), in McKinsey (1952, Theorems 2.6 a 2.7), and in many other publications. □

Note that the common value of numbers C_1 and C_2 ensured by Theorem 12.3 is called the *value of the game*. We denote this by C. The game is *fair* if $C = 0$.

If we add a fixed amount M to each element of the payoff matrix $\boldsymbol{A}$, we can show that the optimal strategies are the same and the value will be $C + M$ instead of C. This result is used in the next section.

12.4 SOLUTION OF MATRIX GAMES

Consider the game with the payoff matrix (12.6). Let $\boldsymbol{u} = (u_1, \ldots, u_m)'$ be the optimal strategy of player A, which we are looking for. Let C be the value of the game. Optimal strategy $\boldsymbol{u}$ ensures to player A an expected prize of at least C for every strategy selected by player B. This is especially true for every pure strategy of B. Then we have

$$\begin{aligned} a_{11}u_1 + \cdots + a_{m1}u_m &\geq C, \\ \cdots\cdots\cdots\cdots\cdots\cdots \\ a_{1n}u_1 + \cdots + a_{mn}u_m &\geq C. \end{aligned} \tag{12.11}$$

We have two possible procedures that will allow us to continue.

Procedure 1. Formally define $u_{m+1} = C$ and take into account the fact that $u_1, \ldots, u_m$ are probabilities. We want to

$$\text{maximize} \quad u_{m+1} \tag{12.12}$$

under constraints

$$\begin{aligned}
a_{11}u_1 + \cdots + a_{m1}u_m - u_{m+1} &\geq 0, \\
\cdots\cdots\cdots\cdots\cdots\cdots\cdots\cdots& \\
a_{1n}u_1 + \cdots + a_{mn}u_m - u_{m+1} &\geq 0, \\
u_1 + \cdots + u_m &= 1, \\
u_i &\geq 0 \quad \text{for} \quad i = 1, \ldots, m.
\end{aligned} \tag{12.13}$$

This is a problem of linear programming, the solution of which will be for a given matrix $\boldsymbol{A}$ solved by a computer. If you are really going to use a computer, do not forget that the computer in many cases demands all coefficients of the objective function, including those that are equal to zero. It does not suffice to pose (12.12); we must want to

$$\text{Maximize} \quad 0 \times u_1 + 0 \times u_2 + \cdots + 0 \times u_m + 1 \times u_{m+1},$$

or

$$\text{Minimize} \quad 0 \times u_1 + 0 \times u_2 + \cdots + 0 \times u_m - 1 \times u_{m+1}$$

(which is, of course, the same). The results given by the computer are not only the probabilities $u_1, \ldots, u_m$, which give the optimal strategy of player A and the value of the game u_{m+1}, but, as a rule, also solution of the so-called *dual problem*, which in our case is fortunately just the optimal strategy $v_1, \ldots, v_n$ of player B.

Since the matrix notation is more transparent, we introduce $\mathbf{1}$ as the symbol for the column vector such that all its components are equal to 1. The number of the components is not written explicitly, if it is clear. Then problems (12.12) and (12.13) can be written in the form

$$\text{Maximize } u_{m+1} \text{ under constraints } \boldsymbol{A}'\boldsymbol{u} - \mathbf{1}u_{m+1} \geq \mathbf{0}, \quad \boldsymbol{u}'\mathbf{1} = 1, \quad \boldsymbol{u} \geq \mathbf{0}. \tag{12.14}$$

In some cases there exist several optimal strategies. The computer usually writes only one of them. The others can be determined from symmetry. We illustrate this method later in concrete examples.

Procedure 2. Assume that $C > 0$. If this condition is not satisfied (or if we are not quite sure about it), then we add to each element a_{ij} of the payoff matrix $\boldsymbol{A}$ a sufficiently large number M such that all elements of the new matrix are positive. This ensures that the value of the new matrix game is positive. As we know, optimal strategies remain the same. Define

$$t_i = \frac{u_i}{C}, \qquad i = 1, \ldots, m.$$

Then (12.11) and constraints concerning nonnegativity imply

$$\begin{aligned}
a_{11}t_1 + \cdots + a_{m1}t_m &\geq 1, \\
\cdots\cdots\cdots\cdots\cdots\cdots& \\
a_{1n}t_1 + \cdots + a_{mn}t_m &\geq 1, \\
t_i &\geq 0 \quad \text{for} \quad i = 1, \ldots, m.
\end{aligned} \tag{12.15}$$

The relation $t_1 + \cdots + t_m = 1$ is changed to $t_1 + \cdots + t_m = 1/C$. The value of the game is maximal in the case that $1/C$ is minimal. Thus we have a linear programming problem

$$\text{Minimize} \quad t_1 + \cdots + t_m \tag{12.16}$$

under system of constraints (12.15). As soon as we find a solution, the value of the game is determined from

$$C = \frac{1}{\sum_{i=1}^{m} t_i} \tag{12.17}$$

and optimal strategy of player A is $\boldsymbol{u} = (u_1, \ldots, u_m)'$, where

$$u_i = \frac{t_i}{\sum_{j=1}^{m} t_j}, \qquad i = 1, \ldots, m. \tag{12.18}$$

If we added at the beginning a constant M to all elements a_{ij}, we would obtain the value of the original game if we were to subtract M from the value C introduced in (12.17).

Define $\boldsymbol{t} = (t_1, \ldots, t_m)'$. Then (12.16) and (12.15) can be written as the problem

$$\text{Minimize } \boldsymbol{t}'\mathbf{1} \text{ under constraints } \boldsymbol{A}'\boldsymbol{t} \geq \mathbf{1},\ \boldsymbol{t} \geq \mathbf{0}. \tag{12.19}$$

12.5 SOLUTION OF 2 × 2 GAMES

The simplest case is

$$\boldsymbol{A} = \begin{pmatrix} a_{11} & a_{12} \\ a_{21} & a_{22} \end{pmatrix}. \tag{12.20}$$

It is presumably possible to calculate here the optimal strategy explicitly without using strong means of linear programming.

First, it is necessary to determine whether $\boldsymbol{A}$ has a saddle point. If this is the case, everything is clear and no calculations are needed. Assume that $\boldsymbol{A}$ has no saddle point. Let $\boldsymbol{u} = (u_1, u_2)'$ be the optimal strategy of player A. We know that

$$u_1 + u_2 = 1. \tag{12.21}$$

It can be proved that under our assumptions player A reaches expected prize equal to the value of the game even if B selects each of B's two pure strategies. This yields

$$\begin{aligned} a_{11}u_1 + a_{21}u_2 &= C, \\ a_{12}u_1 + a_{22}u_2 &= C. \end{aligned} \tag{12.22}$$

We calculate from (12.21) and (12.22)

$$u_1 = \frac{a_{22} - a_{21}}{a_{11} + a_{22} - a_{12} - a_{21}}, \qquad u_2 = 1 - u_1. \tag{12.23}$$

The value of the game follows from (12.22). However, before using formula (12.23), check once more that $\boldsymbol{A}$ has no saddle point.

As an example, consider the *game of matching pennies* with the payoff matrix (12.9). The payoff matrix has no saddle point, and inserting terms into (12.23), we get $u_1 = u_2 = \frac{1}{2}$. The value of the game $C = 0$ follows from (12.22). From symmetry we can see that the optimal strategy of player B is also $v_1 = v_2 = \frac{1}{2}$. The game is fair.

In the following sections of this chapter we find optimal strategies for some less known trivial games.

12.6 TWO-FINGER MORA

It seems that the game named mora is quite popular in some countries. For example, in the last volume of the famous novel *The Count of Monte-Cristo* by Alexandre Dumas (1802–1870), we read: "Peppino had much time. He played *mora* with facchinas, he lost three goldens and to comfort himself he drank a bottle of orviet vine." In explanations we find: *mora* (Ital.) — a game in which a player tries to guess how many fingers the opponent is showing.

The variant of mora, which we consider in this section, has the following rules. Each player shows either one or two fingers and guesses how many fingers the other player shows. If both players guessed correctly or if both of them failed, the game is a tie and both obtain zero prize. If only one player guesses correctly, that player gets the amount equal to the sum of fingers that were shown by both players together.

The strategy of each player has two components. The first component gives the number of fingers which the player shows and the second component is the guess of the number of fingers which the other player shows. Then the payoff matrix is

$$
\begin{array}{cc}
 & \text{Player B} \\
 & \begin{array}{cccc} (1,1) & (1,2) & (2,1) & (2,2) \end{array} \\
\text{Player A} \begin{array}{c} (1,1) \\ (1,2) \\ (2,1) \\ (2,2) \end{array} &
\begin{pmatrix} 0 & 2 & -3 & 0 \\ -2 & 0 & 0 & 3 \\ 3 & 0 & 0 & -4 \\ 0 & -3 & 4 & 0 \end{pmatrix}.
\end{array}
$$

This is the *skew–symmetric matrix* ($\boldsymbol{A}' = -\boldsymbol{A}$). A game with the skew–symmetric payoff matrix is called a *symmetric game*. It is proved that the value of every symmetric game is zero (see Karlin 1959, Sect. 2.6). Moreover, it is proved that a symmetric game with a payoff matrix having an even number of rows has an optimal strategy such that it is not completely mixed (Karlin 1959, Sect. 2.7, Exercise 20). Practically, this means that in our case we find at least one pure strategy of player A that A must not or need not use in A's mixed optimal strategy. Analogous assertion also holds for player B. Using a computer, we get the following optimal strategies:

$$\boldsymbol{u} = \boldsymbol{v} = (0; 0.6; 0.4; 0)'.$$

The value of the game was calculated as $C = 0$, which is already known from our previous considerations.

12.7 THREE-FINGER MORA

The term *mora*, if used alone, generally implies the game three finger mora. The rules are analogous to those from the previous section. Each player shows one, two, or three fingers. In the same time both player guess how many fingers the opponent is showing. If both players guessed either correctly or incorrectly, the payoff would be zero. If only one player guessed correctly, that player gets an amount equal to the sum of the fingers shown by both players together. Every strategy can be written as a vector with two components. The first component is the number of fingers the player shows. The second component is the number of fingers the player guesses the other is showing. Thus we get the payoff matrix

$$\begin{array}{c} \\ (1,1)\\ (1,2)\\ (1,3)\\ (2,1)\\ (2,2)\\ (2,3)\\ (3,1)\\ (3,2)\\ (3,3) \end{array}
\begin{array}{c} \begin{array}{ccccccccc} (1,1) & (1,2) & (1,3) & (2,1) & (2,2) & (2,3) & (3,1) & (3,2) & (3,3) \end{array} \\
\begin{pmatrix} 0 & 2 & 2 & -3 & 0 & 0 & -4 & 0 & 0 \\ -2 & 0 & 0 & 0 & 3 & 3 & -4 & 0 & 0 \\ -2 & 0 & 0 & -3 & 0 & 0 & 0 & 4 & 4 \\ 3 & 0 & 3 & 0 & -4 & 0 & 0 & -5 & 0 \\ 0 & -3 & 0 & 4 & 0 & 4 & 0 & -5 & 0 \\ 0 & -3 & 0 & 0 & -4 & 0 & 5 & 0 & 5 \\ 4 & 4 & 0 & 0 & 0 & -5 & 0 & 0 & -6 \\ 0 & 0 & -4 & 5 & 5 & 0 & 0 & 0 & -6 \end{pmatrix} \end{array}.$$

This is again the symmetric game. As for its numerical solution, Williams (1954) writes: "Now we would need a glimpse of geniality, since the direct computation is hopeless." Then he introduces the solution

$$\boldsymbol{u} = \boldsymbol{v} = (0, 0, \frac{5}{12}, 0, \frac{4}{12}, 0, \frac{3}{12}, 0, 0,)'.$$

The same result we get using a computer in a standard way. The value of the game is zero, since the game is symmetric.

12.8 PROBLEM OF COLONEL BLOTTO

The theory of games is, in fact, a theory of conflict situations. No wonder that the army also seized it (see Aškenazy 1961). The problem of Colonel Blotto seems to be the best known and the most popular one. On the other hand, it is so simple that it is not introduced in Aškenazy's book.

Colonel Blotto commands six regiments. His opponent has five regiments. They fight in two locations that are rather far away. Assume that the fight has the following rules:

- Each commander can divide his formations into only regiments.

- Each commander must send at least one regiment to each location.
- If both opponents have the same number of regiments at a given location, it is a draw and both sides are completely destroyed there.
- If one opponent has more regiments than the other at a given location, he destroys the enemy and has no loss.
- The payoff is the number of destroyed regiments of the enemy after subtracting the first opponent's own destroyed regiments.

The strategy of each opponent can be given as a vector with two components. The first component is the number of regiments sent to fight to the first location and the second component is the number of regiments sent to the other location. This gives the payoff matrix

		Opponent			
		(4, 1)	(3, 2)	(2, 3)	(1, 4)
Colonel Blotto	(5, 1)	4	2	1	0
	(4, 2)	1	3	0	−1
	(3, 3)	−2	2	2	−2
	(2, 4)	−1	0	3	1
	(1, 5)	0	1	2	4

Using a computer, we deduce that Colonel Blotto has the optimal strategy $\boldsymbol{u} = (\frac{4}{9}, 0, \frac{1}{9}, 0, \frac{4}{9})'$. The optimal strategy of his opponent is $\boldsymbol{v}_1 = (\frac{3}{90}, \frac{48}{90}, \frac{32}{90}, \frac{7}{90})'$. Considering certain symmetry of the game it is obvious that other optimal strategies of the opponent are $\boldsymbol{v}_2 = (\frac{7}{90}, \frac{32}{90}, \frac{48}{90}, \frac{3}{90})'$ and $\boldsymbol{v} = \frac{1}{2}(\boldsymbol{v}_1 + \boldsymbol{v}_2) = (\frac{1}{18}, \frac{8}{18}, \frac{8}{18}, \frac{1}{18})'$. The value of the game is $3\frac{5}{9}$. A modified problem of Colonel Blotto is introduced in Karlin (1959). Williams (1954) considers a continuation of the battle where the regiment remaining alive continue to fight.

12.9 SCISSORS— PAPER — STONE

This game is very popular. Player A shows either a fist (representning a stone), or three fingers (representning scissors) or all five fingers (representing paper). Player B does the same. Stone wins over scissors (they will be blunt), scissors win over paper (they cut it), and paper wins over stone (paper wraps stone). The payoff matrix is

		Player B		
		Stone	Scissors	Paper
Player A	Stone	0	1	−1
	Scissors	−1	0	1
	Paper	1	−1	0

Calculations confirm the known fact that the optimal strategies are $\boldsymbol{u} = \boldsymbol{v} = (\frac{1}{3}, \frac{1}{3}, \frac{1}{3})'$. The value of the game is zero, since the game is symmetric. The payoff matrix has odd number of rows. Therefore, the fact that the optimal strategy is completely mixed does not contradict the theory given above.

12.10 BIRTHDAY

The following nice problem is introduced in Williams (1954). The husband returns home in the evening and he suddenly realize that his wife may have her birthday today. However, he is not quite sure about it. He must solve the following problem to determine whether he should buy a bunch of flowers today. If he does not buy the flowers and his wife has no birthday today, everything is okay. He thinks that his payoff is zero in this case. If his wife has a birthday today and he comes without flowers, his situation is terrible. He estimates his payoff by the number -10. His opinion about the remaining two cases follows from the payoff matrix:

	No birthday	Birthday
No flowers	0	-10
Flowers	1	1.5

Inserting terms into (12.23) we get $u_1 = \frac{1}{21} = 0.048$, $u_2 = 1 - u_1 = \frac{20}{21} = 0.952$. The optimal solution of the problem seems to be buying a bunch of flowers with probability $\frac{20}{21}$ today.

However, this solution is incorrect. The payoff matrix has the saddle point $a_{21} = 1$. Have you also forgotten to check it? Hence the optimal strategy is to buy flowers in any case. A man who has been married for several years would know this without any calculation.

References

1. Anděl, J. (1978). *Mathematical Statistics* (in Czech). SNTL/ALFA, Prague.

2. Anděl, J. (1985/86). How to find a straight line going through several points (in Czech). *Rozhledy matematicko fyzikální* **64**:185–190.

3. Anděl, J. (1988a). To command, or to vote? (in Czech). *Věda a technika mládeži* **XLII** (10):30–32.

4. Anděl, J. (1988b). Why the records are so rare (in Czech). *Věda a technika mládeži* **XLII** (4):28–32.

5. Anděl, J. (1988c). Dice and transitivity (in Czech). *Rozhledy matematicko fyzikální* **67**:309–313.

6. Anděl, J. (1990). Report on a business trip abroad (in Czech). *Inform. Bull. Czech Statist. Soc.* (3), 3.

7. Anděl, J. (1990/91). Principle of reflection and reliability of service (in Czech). *Rozhledy matematicko fyzikální* **69**:54–60.

8. Anděl, J. (1991a). Christmas inequality. *Inform. Bull. Czech Statist. Soc.* (Engl. issue) 8–10.

9. Anděl, J. (1991b). Christmas inequality (in Czech). *Inform. Bull. Czech Statist. Soc.* (4), 1–4.

10. Anděl, J. (1993). How to win a million (in Czech). *Inform. Bull. Czech Statist. Soc.* (4), 1–5.

11. Anděl, J., and Zvára, K. (1980/81). On quadratic equations (in Czech). *Rozhledy matematicko fyzikální*, **59**:339–344.

12. Aškenazy, V. O. (1961). *Application of Theory of Games in Military Problems* (in Russian). Izd. Sovetskoje radio, Moscow. (A collection of translated English papers.)

13. Austin, J. D. (1982). Overbooking airline flights. *The Math. Teacher* **75**:221–223.

14. Banjevič, D. (1991). Inequalities on the expectation and the variance of a function of a random variable. Unpublished manuscript.

15. Banjevič, D., and Bratičevič, D. (1983). Note on dispersion of X^α. *Publ. Inst. Math.* (new series) **33**(47):23–28.

16. Barlow, R. E., Hunter, L. C., and Proschan, F. (1963). Optimum redundancy, when components are subject to two kinds of failure. *J. Soc. Indust. Appl. Math.* **11**:64–73.

17. Bernardo, J. M., and Smith, A. F. M. (1994). *Bayesian Theory.* Wiley, New York.

18. Bernoulli, J. (1713). *Ars Conjectandi.* Reprinted in *Die Werke von Jakob Bernoulli*, Vol. 3 (1975), Birkhäuser, Basel.

19. Bharucha-Reid, A. T., Sambandham, M. (1986). *Random Polynomials.* Academic Press, Orlando.

20. Billingsley, P. (1986). *Probability and Measure.* Wiley, New York.

21. Blom, G., Holst, L., and Sandell, D. (1994). *Problems and Snapshots from the World of Probability.* Springer, New York.

22. Bloomfield, P., and Steiger W. L. (1983). *Least Absolute Deviations.* Birkhäuser, Boston.

23. Boas, R. P. (1981). Snowfall and elephants, pop bottles and π. *The Math. Teacher* **74**:49–55.

24. Boer, H. (1993). AIDS — welche Aussagekraft hat ein "positives" Test-Ergebnis? *Stochastik in der Schule* **13**:2–12.

25. Bühlmann, H. (1998). Mathematische Paradigmen in der Finanzwirtschaft. *Elem. Math.* **53**:159–176.

26. Carroll, L. (1893). *Pillow Problems.* 4th ed. (1895), Problem 58, pp. 14, 25, 83–84; reprinted (in *A Tangled Tale*), Dover Publications, New York 1958.

27. Chan, B. (1984). Moving on. *The Actuary* **18**:4–5.

28. Chan, W.-S. (1996). Mang kung dice game. *Teaching Statist.* **18**:42–44.

29. Coyle, C. A., and Wang, C. (1993). Wanna bet? On gambling strategies that may or may not work in a casino. *Am. Statist.* **47**:108–111.

30. Cramér, H. (1946). *Mathematical Methods of Statistics.* Princeton Univ. Press, Princeton, NJ.

31. Croucher, J. S. (1985). Changing the rules of tennis: Who has the advantage? *Teaching Statist.* **7**:82–84.

32. Deshpande, M. N. (1992). A question on random variables. *Teaching Statist.*, **14**:9.

33. Dorfman, R. (1943). The detection of defective members of large populations. *Ann. Math. Statist.* **14**:436–440.

34. Dubins, L. E., and Savage, L. (1965). *Inequalities for Stochastic Processes: How to Gamble If You Must.* Dover Publications, New York.

35. Dupačová, J. (1982). *Linear Programming* (in Czech). SPN, Prague.

36. Eisenberg, B., and Sullivan, R. (1996). Random triangles in n dimensions. *Am. Math. Monthly* **103**:308–318.

37. Engel, A. (1993). The computer solves the three tower problem. *Am. Math. Monthly* **100**:62–64.

38. Engel, J. (1996). Das Achensee-Paradoxon. *Stochastik in der Schule* **16**:3–10.

39. Everitt, B. S., (1999). *Chance Rules. An Informal Guide to Probability, Risk, and Statistics.* Copernikus, Springer, New York.

40. Farahmand, K. (1991). Real zeros of random algebraic polynomials. *Proc. Am. Math. Soc.* **113**:1077–1084.

41. Feinerman, R. (1976). An ancient unfair game. *Am. Math. Monthly* **83**:623–625.

42. Feller, W. (1968). *An Introduction to Probability Theory and Its Applications*, Vols. I and II. Wiley, New York.

43. Ferguson, T. S., and Hardwick, J. P. (1989). Stopping rules for proofreading. *J. Appl. Probab.* **26**:304–313.

44. Field, D. A. (1978). Investigating mathematical models. *Am. Math. Monthly* **85**:196–197.

45. Fisher, R. A. (1934). *Statistical Methods for Research Workers.* Oliver and Boyd, Edinburgh.

46. Freeman, P. R. (1983). The secretary problem and its extensions: A review. *Internat. Statist. Rev.* **51**:189–206.

47. Gaver, D. P., and Powell, B. A. (1971). Variability in round-trip times for an elevator car during up-peak. *Transportation Sci.* **5**:169.

48. Glick, N. (1978). Breaking records and breaking boards. *Am. Math. Monthly*, **85**:2–26.

49. Glickman, L. (1991). Isaac Newton—the modern consultant. *Teaching Statist.* **13**:66–67.

50. Gordon, H. (1997). *Discrete Probability.* Springer, New York.

51. Gottinger, H. W. (1980). *Elements of Statistical Analysis.* De Gruyter, Berlin–New York.

52. Guy, R. K. (1993). There are three times as many obtuse-angled triangles as there are acute-angled ones. *Math. Mag.* **66**:175–179.

53. Hader, R. J. (1967). Random roomate pairing of negro and white students. *Am. Statist.* **21** (5):24–26.

54. Hájek J., and Šidák Z. (1967). *Theory of Rank Tests.* Academia, Prague.

55. Hald, A. (1990). *A History of Probability & Statistics and Their Applications Before 1750.* Wiley, New York.

56. Hall, G. R. (1982). Acute triangles in n-ball. *J. Appl. Probab.*, 19:712–715.

57. Hendricks, W. J. (1972). The stationary distribution of an interesting Markov chain. *J. Appl. Probab.* **9**:231–233.

58. Henningsen, J. (1984). An activity for predicting performances in the 1984 Summer Olympics. *The Math. Teacher* **77**:338–341.

59. Jaglom, A. M., Jaglom, I. M. (1954). *Non-elementary problems in elementary presentation* (in Russian). Gos. izd. techniko-teoretičeskoj literatury, Moscow.

60. Karlin, S. (1959). *Mathematical Methods and Theory in Games, Programming, and Economics.* Pergamon Press, London.

61. Kaucký, J. (1975). *Combinatorial Identities* (in Czech). Veda, Bratislava.

62. Kendall, M. G., and Stuart A. (1973). *The Advanced Theory of Statistics* II, 3rd ed. Griffin, London.

63. Kirschenhofer, P., and Prodinger, H. (1994). The higher moments of the number of returns of a simple random walk. *Adv. Appl. Probab.* **26**:561–563.

64. Kolmogorov, A. N. (1950). Unbiased estimates (in Russian). *Izv. Akad. Nauk SSSR, Ser. Mat.* **14**:303–326.

65. Komenda, S. (1997). From a fairytale to a fairytale and then back or looking for Cinderella as object of managenent (in Czech). Preprint.

66. Kulichová, J. (1994). Reliability of serial-paralel systems with two dual types of failures (in Czech). Diploma work, Charles Univ., Prague.

67. Lam, K., Leung, M.-Y., and Siu, M.-K. (1983). Self-organizing files. *UMAP J.* **4**:51–84.

68. Li, H. C. (1988). The exact probability that the roots of quadratic, cubic, and quartic equations are all real if the equation coefficients are random. *Commun. Statist. – Th. Meth.* **17**:395–409.

69. Liu, F. (1988). A note on probabilities in proofreading. *Am. Math. Monthly* **95**:854.

70. Loève, M. (1955). *Probability Theory.* Van Nostrand, Princeton.

71. Lozansky, E., and Rousseau, C. (1996). *Winning Solutions.* Springer, New York.

72. Ludvíková, J. (1997). Georg Pick (1859–1942), life and main orientations of his activities (in Czech). Diploma work, Charles Univ.

73. Mačák, K. (1997). *Origins of the Probability Theory* (in Czech). Prometheus, Prague.

74. McKinsey, J. C. C. (1952). *Introduction to the Theory of Games.* McGraw-Hill, New York.

75. Mead, R. (1992). Statistical games 2—medical diagnosis. *Teaching Statist.*, **14**:12–16.

76. *Meteorological Observations in Prague — Klementinum)* (1976), Vols. I (1775–1900) and II (1901–1975) (in Czech). Institute of Hydrometeorology, Prague.

77. Mishra, M. N., Nayak, N. N., Pattanayak, S. (1982). Strong result for real zeros of random polynomials. *Pacific J. Math.* **103**:502–523.

78. Mishra, M. N., Nayak, N. N., Pattanayak, S. (1983). Lower bound for the number of real roots of a random algebraic polynomial. *J. Austr. Math. Soc.* **A 35**:18–27.

79. Moore, E. F., and Shannon C. E. (1956). Reliable circuits using less reliable relays. *J. Franklin Inst.* **262**:191–208, 281–297.

80. Mosteller, F. (1965). *Fifty Challenging Problems in Probability with Solutions.* Addison-Wesley, Reading, MA.

81. Natanson, I. P. (1957). *Theory of Functions of Real Variable* (in Russian). Gos. izd. techniko-teor. lit., Moscow.

82. Netuka, I. (1999). Georg Pick — mathematical colleague of Albert Einstein in Prague (in Czech). *Pokroky matematiky, fyziky a astronomie* **44**:227–232.

83. O'Brien, T. (1985). Fixed points and exam taking strategy. *Am. Math. Monthly* **92**:659–661.

84. Palacios, J. L. (1999). The ruin problem via electric networks. *Am. Statist.* **53**:67–70.

85. Pick, G. (1899). Geometrisches zur Zahlenlehre. *Sitzungsberichte des deutschen naturwissenschaftlich-medicinischen Vereines f. Böhmen "Lotos"* **19**(47).

86. Pinl, M., and Dick, A. (1974). Kollegen in einer dunklen Zeit. Schluß. *Jber. Deutsch. Math.-Verein* **75**:166–208.

87. Pólya, G. (1976). Probabilities in proofreading. *Am. Math. Monthly* **83**:42.

88. Portnoy, S. (1994). A Lewis Carroll pillow problem: Probability of an obtuse triangle. *Statist. Sci.* **9**:279–284.

89. Rabinowitz, S. (Ed.) (1992). *Index to Mathematical Problems 1980–1984.* MathPro Press, Westford, MA.

90. Randall, K. (1984). Quick Mac attack. *The Math. Teacher* **77**:465, 487.

91. Rao, C. R. (1973). *Linear Statistical Inference and Its Applications*, 2nd ed. Wiley, New York.

92. Read, R. C. (1966). A type of "gambler's ruin" problem. *Am. Math. Monthly* **73**:177–179.

93. Rektorys, K. et al. (1995). *Review of Applied Mathematics,* 6th ed., Vols. I and II (in Czech). Prometheus, Prague.

94. Rényi, A. (1962). Théorie des élements saillants d'une suite d'observations. *Colloquium on Combinatorial Methods in Probability Theory*, pp. 104–117, Matematisk Institut, Aarhus Universitet, Denmark.

95. Rényi, A. (1970). *Probability Theory.* Akadémiai Kiadó, Budapest.

96. Rhodius, A. (1991). Zur Anwendung einiger Funktionalgleichungen in der Wahrscheinlichkeitsrechnung. *Mathematik in der Schule* **29**:793–798.

97. Ridenhour, J. R., and Woodward, E. (1984). The probability of winning in McDonald's Star Raiders contest. *The Math. Teacher* **77**:124–128.

98. Riordan, J. (1968). *Combinatorial Identities.* Wiley, New York.

99. Rosen, M. I. (1995). Niels Hendrik Abel and equations of the fifth degree. *Am. Math. Monthly* **102**:495–505.

100. Ross, S. M. (1994). Comment to Christense, R., and Utts, J. (1992). 'Bayesian resolution of the exchange paradox', Amer. Statist. **47**:311, *Amer. Statist.* **48**:267.

101. Rudin, W. (1974). *Real and Complex Analysis.* McGraw-Hill, New York.

102. Sambandham, M., Thangarai, V., Bharucha-Reid, A. T. (1983). On the variance of the number of real roots of random algebraic polynomials. *Stoch. Anal. Appl.* **1**:215–238.

103. Sankaranarayanan, G. (1979). On the theory of random polynomials. *Math. Student* **47**:172–192.

104. Shenker, M. (1981). The mean number of real roots for one class of random polynomials. *Ann. Prob.* **9**:510–512.

105. Sierpiǹski, W. (1956). *O rozwiazywaniu równaǹ w liczbach całkowitych.* Paǹstwowe wydawnictwo naukowe, Warsaw.

106. Smith, D. E. (1929). *A Source Book in Mathematics.* McGraw-Hill, New York (new ed. 1959).

107. Štěpán, J. (1987). *Probability Theory* (in Czech). Academia, Prague.

108. Stigler, S. M. (1986). Laplace's 1774 memoir on inverse probability. *Statist. Sci.* **1**:359–378.

109. Stirzaker, D. (1994). *Elementary Probability.* Cambridge Univ. Press, Cambridge, UK.

110. Swift, J. (1983). Challenges for enriching the curriculum: Statistics and probability. *The Math. Teacher* **76**:268–269.

111. Székely, G. J. (1986). *Paradoxes in Probability Theory and Mathematical Statistics.* Akadémiai Kiadó, Budapest.

112. Tata, M. N. (1969). On outstanding values in a sequence of random variables. *Z. Wahrsch. Verw. Gebiete* **12**:9–20.

113. Tucker, A. (1984). *Applied Combinatorics.* Wiley, New York.

114. Valfiš, A. Z. (1952). *Uravnenije Pellja.* Izd. Akad. Nauk Gruzinskoj SSR, Tbilisi.

115. Vervaat, W. (1973). Limit theorems for records from discrete distributions. *Stochastic Process. Appl.* **1**:317–334.

116. Wang, C. (1993). *Sense and Nonsense of Statistical Inference.* Marcel Dekker, New York.

117. Williams, J. D. (1954). *The Compleat Strategyst Being a Primer on the Theory of Games of Strategy.* McGraw-Hill, New York.

118. Wolfram, S. (1991). *Mathematica.* Addison-Wesley, New York.

119. Wong, E. (1964). A linear search problem. *SIAM Rev.* **6**:168–174.

120. Woolhouse, W. S. B. (1886). Solution by the proposer to problem 1350. In *Mathematical Questions, with their Solutions from the Educational Times*, Vol. 1, pp. 49–51, C. F. Hodgson and Son, London.

Topic Index

Author Index

WILEY SERIES IN PROBABILITY AND STATISTICS

ESTABLISHED BY WALTER A. SHEWHART AND SAMUEL S. WILKS

Probability and Statistics Section

*ANDERSON · The Statistical Analysis of Time Series

ARNOLD, BALAKRISHNAN, and NAGARAJA · A First Course in Order Statistics

ARNOLD, BALAKRISHNAN, and NAGARAJA · Records

BACCELLI, COHEN, OLSDER, and QUADRAT · Synchronization and Linearity: An Algebra for Discrete Event Systems

BARNETT · Comparative Statistical Inference, *Third Edition*

BASILEVSKY · Statistical Factor Analysis and Related Methods: Theory and Applications

BERNARDO and SMITH · Bayesian Statistical Concepts and Theory

BILLINGSLEY · Convergence of Probability Measures, *Second Edition*

BOROVKOV · Asymptotic Methods in Queuing Theory

BOROVKOV · Ergodicity and Stability of Stochastic Processes

BRANDT, FRANKEN, and LISEK · Stationary Stochastic Models

CAINES · Linear Stochastic Systems

CAIROLI and DALANG · Sequential Stochastic Optimization

CONSTANTINE · Combinatorial Theory and Statistical Design

COOK · Regression Graphics

COVER and THOMAS · Elements of Information Theory

CSÖRGŐ and HORVÁTH · Weighted Approximations in Probability Statistics

CSÖRGŐ and HORVÁTH · Limit Theorems in Change Point Analysis

*DANIEL · Fitting Equations to Data: Computer Analysis of Multifactor Data, *Second Edition*

DETTE and STUDDEN · The Theory of Canonical Moments with Applications in Statistics, Probability, and Analysis

DEY and MUKERJEE · Fractional Factorial Plans

*DOOB · Stochastic Processes

DRYDEN and MARDIA · Statistical Shape Analysis

DUPUIS and ELLIS · A Weak Convergence Approach to the Theory of Large Deviations

ETHIER and KURTZ · Markov Processes: Characterization and Convergence

FELLER · An Introduction to Probability Theory and Its Applications, Volume I, *Third Edition,* Revised; Volume II, *Second Edition*

FULLER · Introduction to Statistical Time Series, *Second Edition*

FULLER · Measurement Error Models

GHOSH, MUKHOPADHYAY, and SEN · Sequential Estimation

GIFI · Nonlinear Multivariate Analysis

GUTTORP · Statistical Inference for Branching Processes

HALL · Introduction to the Theory of Coverage Processes

HAMPEL · Robust Statistics: The Approach Based on Influence Functions

HANNAN and DEISTLER · The Statistical Theory of Linear Systems

HUBER · Robust Statistics

*Now available in a lower priced paperback edition in the Wiley Classics Library.

Probability and Statistics (Continued)

HUSKOVA, BERAN, and DUPAC · Collected Works of Jaroslav Hajek—with Commentary
IMAN and CONOVER · A Modern Approach to Statistics
JUREK and MASON · Operator-Limit Distributions in Probability Theory
KASS and VOS · Geometrical Foundations of Asymptotic Inference
KAUFMAN and ROUSSEEUW · Finding Groups in Data: An Introduction to Cluster Analysis
KELLY · Probability, Statistics, and Optimization
KENDALL, BARDEN, CARNE, and LE · Shape and Shape Theory
LINDVALL · Lectures on the Coupling Method
MANTON, WOODBURY, and TOLLEY · Statistical Applications Using Fuzzy Sets
MORGENTHALER and TUKEY · Configural Polysampling: A Route to Practical Robustness
MUIRHEAD · Aspects of Multivariate Statistical Theory
OLIVER and SMITH · Influence Diagrams, Belief Nets and Decision Analysis
*PARZEN · Modern Probability Theory and Its Applications
PEÑA, TIAO, and TSAY · A Course in Time Series Analysis
PRESS · Bayesian Statistics: Principles, Models, and Applications
PUKELSHEIM · Optimal Experimental Design
RAO · Asymptotic Theory of Statistical Inference
RAO · Linear Statistical Inference and Its Applications, *Second Edition*
RAO and SHANBHAG · Choquet-Deny Type Functional Equations with Applications to Stochastic Models
ROBERTSON, WRIGHT, and DYKSTRA · Order Restricted Statistical Inference
ROGERS and WILLIAMS · Diffusions, Markov Processes, and Martingales, Volume I: Foundations, *Second Edition;* Volume II: Îto Calculus
RUBINSTEIN and SHAPIRO · Discrete Event Systems: Sensitivity Analysis and Stochastic Optimization by the Score Function Method
RUZSA and SZEKELY · Algebraic Probability Theory
SCHEFFE · The Analysis of Variance
SEBER · Linear Regression Analysis
SEBER · Multivariate Observations
SEBER and WILD · Nonlinear Regression
SERFLING · Approximation Theorems of Mathematical Statistics
SHORACK and WELLNER · Empirical Processes with Applications to Statistics
SMALL and McLEISH · Hilbert Space Methods in Probability and Statistical Inference
STAPLETON · Linear Statistical Models
STAUDTE and SHEATHER · Robust Estimation and Testing
STOYANOV · Counterexamples in Probability
TANAKA · Time Series Analysis: Nonstationary and Noninvertible Distribution Theory
THOMPSON and SEBER · Adaptive Sampling
WELSH · Aspects of Statistical Inference
WHITTAKER · Graphical Models in Applied Multivariate Statistics
YANG · The Construction Theory of Denumerable Markov Processes

Applied Probability and Statistics Section

ABRAHAM and LEDOLTER · Statistical Methods for Forecasting
AGRESTI · Analysis of Ordinal Categorical Data
AGRESTI · Categorical Data Analysis

*Now available in a lower priced paperback edition in the Wiley Classics Library.

Applied Probability and Statistics (Continued)

ANDERSON, AUQUIER, HAUCK, OAKES, VANDAELE, and WEISBERG · Statistical Methods for Comparative Studies
*ARTHANARI and DODGE · Mathematical Programming in Statistics
ASMUSSEN · Applied Probability and Queues
*BAILEY · The Elements of Stochastic Processes with Applications to the Natural Sciences
BARNETT and LEWIS · Outliers in Statistical Data, *Third Edition*
BARTHOLOMEW, FORBES, and McLEAN · Statistical Techniques for Manpower Planning, *Second Edition*
BASU and RIGDON · Statistical Methods for the Reliability of Repairable Systems
BATES and WATTS · Nonlinear Regression Analysis and Its Applications
BECHHOFER, SANTNER, and GOLDSMAN · Design and Analysis of Experiments for Statistical Selection, Screening, and Multiple Comparisons
BELSLEY · Conditioning Diagnostics: Collinearity and Weak Data in Regression
BELSLEY, KUH, and WELSCH · Regression Diagnostics: Identifying Influential Data and Sources of Collinearity
BHAT · Elements of Applied Stochastic Processes, *Second Edition*
BHATTACHARYA and WAYMIRE · Stochastic Processes with Applications
BIRKES and DODGE · Alternative Methods of Regression
BLISCHKE AND MURTHY · Reliability: Modeling, Prediction, and Optimization
BLOOMFIELD · Fourier Analysis of Time Series: An Introduction, *Second Edition*
BOLLEN · Structural Equations with Latent Variables
BOULEAU · Numerical Methods for Stochastic Processes
BOX · Bayesian Inference in Statistical Analysis
BOX and DRAPER · Empirical Model-Building and Response Surfaces
*BOX and DRAPER · Evolutionary Operation: A Statistical Method for Process Improvement
BUCKLEW · Large Deviation Techniques in Decision, Simulation, and Estimation
BUNKE and BUNKE · Nonlinear Regression, Functional Relations and Robust Methods: Statistical Methods of Model Building
CHATTERJEE and HADI · Sensitivity Analysis in Linear Regression
CHERNICK · Bootstrap Methods: A Practitioner's Guide
CHILÈS and DELFINER · Geostatistics: Modeling Spatial Uncertainty
CLARKE and DISNEY · Probability and Random Processes: A First Course with Applications, *Second Edition*
*COCHRAN and COX · Experimental Designs, *Second Edition*
CONOVER · Practical Nonparametric Statistics, *Second Edition*
CORNELL · Experiments with Mixtures, Designs, Models, and the Analysis of Mixture Data, *Second Edition*
*COX · Planning of Experiments
CRESSIE · Statistics for Spatial Data, *Revised Edition*
DANIEL · Applications of Statistics to Industrial Experimentation
DAVID · Order Statistics, *Second Edition*
*DEGROOT, FIENBERG, and KADANE · Statistics and the Law
DODGE · Alternative Methods of Regression
DOWDY and WEARDEN · Statistics for Research, *Second Edition*
GALLANT · Nonlinear Statistical Models
GLASSERMAN and YAO · Monotone Structure in Discrete-Event Systems
GNANADESIKAN · Methods for Statistical Data Analysis of Multivariate Observations, *Second Edition*
GOLDSTEIN and LEWIS · Assessment: Problems, Development, and Statistical Issues
GREENWOOD and NIKULIN · A Guide to Chi-Squared Testing
*HAHN · Statistical Models in Engineering

*Now available in a lower priced paperback edition in the Wiley Classics Library.

Applied Probability and Statistics (Continued)

HAHN and MEEKER · Statistical Intervals: A Guide for Practitioners
HAND · Construction and Assessment of Classification Rules
HAND · Discrimination and Classification
HEIBERGER · Computation for the Analysis of Designed Experiments
HEDAYAT and SINHA · Design and Inference in Finite Population Sampling
HINKELMAN and KEMPTHORNE: · Design and Analysis of Experiments, Volume 1: Introduction to Experimental Design
HOAGLIN, MOSTELLER, and TUKEY · Exploratory Approach to Analysis of Variance
HOAGLIN, MOSTELLER, and TUKEY · Exploring Data Tables, Trends and Shapes
HOAGLIN, MOSTELLER, and TUKEY · Understanding Robust and Exploratory Data Analysis
HOCHBERG and TAMHANE · Multiple Comparison Procedures
HOCKING · Methods and Applications of Linear Models: Regression and the Analysis of Variables
HOGG and KLUGMAN · Loss Distributions
HOSMER and LEMESHOW · Applied Logistic Regression, *Second Edition*
HØYLAND and RAUSAND · System Reliability Theory: Models and Statistical Methods
HUBERTY · Applied Discriminant Analysis
JACKSON · A User's Guide to Principle Components
JOHN · Statistical Methods in Engineering and Quality Assurance
JOHNSON · Multivariate Statistical Simulation
JOHNSON and KOTZ · Distributions in Statistics
JOHNSON, KOTZ, and BALAKRISHNAN · Continuous Univariate Distributions, Volume 1, *Second Edition*
JOHNSON, KOTZ, and BALAKRISHNAN · Continuous Univariate Distributions, Volume 2, *Second Edition*
JOHNSON, KOTZ, and BALAKRISHNAN · Discrete Multivariate Distributions
JOHNSON, KOTZ, and KEMP · Univariate Discrete Distributions, *Second Edition*
JUREČKOVÁ and SEN · Robust Statistical Procedures: Aymptotics and Interrelations
KADANE AND SCHUM · A Probabilistic Analysis of the Sacco and Vanzetti Evidence
KELLY · Reversability and Stochastic Networks
KHURI, MATHEW, and SINHA · Statistical Tests for Mixed Linear Models
KLUGMAN, PANJER, and WILLMOT · Loss Models: From Data to Decisions
KLUGMAN, PANJER, and WILLMOT · Solutions Manual to Accompany Loss Models: From Data to Decisions
KOTZ, BALAKRISHNAN, and JOHNSON · Continuous Multivariate Distributions, Volume 1, *Second Edition*
KOVALENKO, KUZNETZOV, and PEGG · Mathematical Theory of Reliability of Time-Dependent Systems with Practical Applications
LAD · Operational Subjective Statistical Methods: A Mathematical, Philosophical, and Historical Introduction
LePAGE and BILLARD · Exploring the Limits of Bootstrap
LINHART and ZUCCHINI · Model Selection
LITTLE and RUBIN · Statistical Analysis with Missing Data
LLOYD · The Statistical Analysis of Categorical Data
MAGNUS and NEUDECKER · Matrix Differential Calculus with Applications in Statistics and Econometrics, *Revised Edition*
MANN, SCHAFER, and SINGPURWALLA · Methods for Statistical Analysis of Reliability and Life Data
McLACHLAN · Discriminant Analysis and Statistical Pattern Recognition
McLACHLAN and KRISHNAN · The EM Algorithm and Extensions
McLACHLAN and PEEL · Finite Mixture Models
MEEKER and ESCOBAR · Statistical Methods for Reliability Data

*Now available in a lower priced paperback edition in the Wiley Classics Library.

Applied Probability and Statistics (Continued)

MONTGOMERY and PECK · Introduction to Linear Regression Analysis, *Second Edition*
MYERS and MONTGOMERY · Response Surface Methodology: Process and Product in Optimization Using Designed Experiments
NELSON · Accelerated Testing, Statistical Models, Test Plans, and Data Analyses
NELSON · Applied Life Data Analysis
OCHI · Applied Probability and Stochastic Processes in Engineering and Physical Sciences
OKABE, BOOTS, and SUGIHARA · Spatial Tesselations: Concepts and Applications of Voronoi Diagrams
PANKRATZ · Forecasting with Dynamic Regression Models
PANKRATZ · Forecasting with Univariate Box-Jenkins Models: Concepts and Cases
PORT · Theoretical Probability for Applications
PUTERMAN · Markov Decision Processes: Discrete Stochastic Dynamic Programming
RACHEV · Probability Metrics and the Stability of Stochastic Models
RÉNYI · A Diary on Information Theory
RIPLEY · Spatial Statistics
RIPLEY · Stochastic Simulation
ROUSSEEUW and LEROY · Robust Regression and Outlier Detection
RUBIN · Multiple Imputation for Nonresponse in Surveys
RUBINSTEIN · Simulation and the Monte Carlo Method
RUBINSTEIN and MELAMED · Modern Simulation and Modeling
RYAN · Statistical Methods for Quality Improvement, *Second Edition*
SCHIMEK · Smoothing and Regression: Approaches, Computation, and Application
SCHUSS · Theory and Applications of Stochastic Differential Equations
SCOTT · Multivariate Density Estimation: Theory, Practice, and Visualization
*SEARLE · Linear Models
SEARLE · Linear Models for Unbalanced Data
SEARLE, CASELLA, and McCULLOCH · Variance Components
SENNOTT · Stochastic Dynamic Programming and the Control of Queueing Systems
STOYAN, KENDALL, and MECKE · Stochastic Geometry and Its Applications, *Second Edition*
STOYAN and STOYAN · Fractals, Random Shapes and Point Fields: Methods of Geometrical Statistics
THOMPSON · Empirical Model Building
THOMPSON · Sampling
THOMPSON · Simulation: A Modeler's Approach
TIJMS · Stochastic Modeling and Analysis: A Computational Approach
TIJMS · Stochastic Models: An Algorithmic Approach
TITTERINGTON, SMITH, and MAKOV · Statistical Analysis of Finite Mixture Distributions
UPTON and FINGLETON · Spatial Data Analysis by Example, Volume 1: Point Pattern and Quantitative Data
UPTON and FINGLETON · Spatial Data Analysis by Example, Volume II: Categorical and Directional Data
VAN RIJCKEVORSEL and DE LEEUW · Component and Correspondence Analysis
VIDAKOVIC · Statistical Modeling by Wavelets
WEISBERG · Applied Linear Regression, *Second Edition*
WESTFALL and YOUNG · Resampling-Based Multiple Testing: Examples and Methods for *p*-Value Adjustment
WHITTLE · Systems in Stochastic Equilibrium
*ZELLNER · An Introduction to Bayesian Inference in Econometrics

*Now available in a lower priced paperback edition in the Wiley Classics Library.

Biostatistics Section

ARMITAGE and DAVID (editors) · Advances in Biometry
BROWN and HOLLANDER · Statistics: A Biomedical Introduction
CHOW and LIU · Design and Analysis of Clinical Trials: Concepts and Methodologies
DUNN and CLARK · Applied Statistics: Analysis of Variance and Regression, *Second Edition*
*ELANDT-JOHNSON and JOHNSON · Survival Models and Data Analysis
*FLEISS · The Design and Analysis of Clinical Experiments
FLEISS · Statistical Methods for Rates and Proportions, *Second Edition*
FLEMING and HARRINGTON · Counting Processes and Survival Analysis
KADANE · Bayesian Methods and Ethics in a Clinical Trial Design
KALBFLEISCH and PRENTICE · The Statistical Analysis of Failure Time Data
LACHIN · Biostatistical Methods: The Assessment of Relative Risks
LANGE, RYAN, BILLARD, BRILLINGER, CONQUEST, and GREENHOUSE · Case Studies in Biometry
LAWLESS · Statistical Models and Methods for Lifetime Data
LEE · Statistical Methods for Survival Data Analysis, *Second Edition*
MALLER and ZHOU · Survival Analysis with Long Term Survivors
McNEIL · Epidemiological Research Methods
McFADDEN · Management of Data in Clinical Trials
*MILLER · Survival Analysis, *Second Edition*
PIANTADOSI · Clinical Trials: A Methodologic Perspective
WOODING · Planning Pharmaceutical Clinical Trials: Basic Statistical Principles
WOOLSON · Statistical Methods for the Analysis of Biomedical Data

Financial Engineering Section

HUNT and KENNEDY · Financial Derivatives in Theory and Practice
ROLSKI, SCHMIDLI, SCHMIDT, and TEUGELS · Stochastic Processes for Insurance and Finance

Texts, References, and Pocketbooks Section

AGRESTI · An Introduction to Categorical Data Analysis
ANDĚL · Mathematics of Chance
ANDERSON · An Introduction to Multivariate Statistical Analysis, *Second Edition*
ANDERSON and LOYNES · The Teaching of Practical Statistics
ARMITAGE and COLTON · Encyclopedia of Biostatistics: Volumes 1 to 6 with Index
BARTOSZYNSKI and NIEWIADOMSKA-BUGAJ · Probability and Statistical Inference
BENDAT and PIERSOL · Random Data: Analysis and Measurement Procedures, *Third Edition*
BERRY, CHALONER, and GEWEKE · Bayesian Analysis in Statistics and Econometrics: Essays in Honor of Arnold Zellner
BHATTACHARYA and JOHNSON · Statistical Concepts and Methods
BILLINGSLEY · Probability and Measure, *Second Edition*
BOX · R. A. Fisher, the Life of a Scientist
BOX, HUNTER, and HUNTER · Statistics for Experimenters: An Introduction to Design, Data Analysis, and Model Building
BOX and LUCEÑO · Statistical Control by Monitoring and Feedback Adjustment
CHATTERJEE and PRICE · Regression Analysis by Example, *Third Edition*

*Now available in a lower priced paperback edition in the Wiley Classics Library.

Texts, References, and Pocketbooks (Continued)

COOK and WEISBERG · Applied Regression Including Computing and Graphics
COOK and WEISBERG · An Introduction to Regression Graphics
COX · A Handbook of Introductory Statistical Methods
DANIEL · Biostatistics: A Foundation for Analysis in the Health Sciences, *Sixth Edition*
DILLON and GOLDSTEIN · Multivariate Analysis: Methods and Applications
*DODGE and ROMIG · Sampling Inspection Tables, *Second Edition*
DRAPER and SMITH · Applied Regression Analysis, *Third Edition*
DUDEWICZ and MISHRA · Modern Mathematical Statistics
DUNN and CLARK · Basic Statistics: A Primer for the Biomedical Sciences, *Third Edition*
EVANS, HASTINGS, and PEACOCK · Statistical Distributions, *Third Edition*
FISHER and VAN BELLE · Biostatistics: A Methodology for the Health Sciences
FREEMAN and SMITH · Aspects of Uncertainty: A Tribute to D. V. Lindley
GROSS and HARRIS · Fundamentals of Queueing Theory, *Third Edition*
HALD · A History of Probability and Statistics and their Applications Before 1750
HALD · A History of Mathematical Statistics from 1750 to 1930
HELLER · MACSYMA for Statisticians
HOEL · Introduction to Mathematical Statistics, *Fifth Edition*
HOLLANDER and WOLFE · Nonparametric Statistical Methods, *Second Edition*
HOSMER and LEMESHOW · Applied Logistic Regression, *Second Edition*
HOSMER and LEMESHOW · Applied Survival Analysis: Regression Modeling of Time to Event Data
JOHNSON and BALAKRISHNAN · Advances in the Theory and Practice of Statistics: A Volume in Honor of Samuel Kotz
JOHNSON and KOTZ (editors) · Leading Personalities in Statistical Sciences: From the Seventeenth Century to the Present
JUDGE, GRIFFITHS, HILL, LÜTKEPOHL, and LEE · The Theory and Practice of Econometrics, *Second Edition*
KHURI · Advanced Calculus with Applications in Statistics
KOTZ and JOHNSON (editors) · Encyclopedia of Statistical Sciences: Volumes 1 to 9 with Index
KOTZ and JOHNSON (editors) · Encyclopedia of Statistical Sciences: Supplement Volume
KOTZ, REED, and BANKS (editors) · Encyclopedia of Statistical Sciences: Update Volume 1
KOTZ, REED, and BANKS (editors) · Encyclopedia of Statistical Sciences: Update Volume 2
LAMPERTI · Probability: A Survey of the Mathematical Theory, *Second Edition*
LARSON · Introduction to Probability Theory and Statistical Inference, *Third Edition*
LE · Applied Categorical Data Analysis
LE · Applied Survival Analysis
MALLOWS · Design, Data, and Analysis by Some Friends of Cuthbert Daniel
MARDIA · The Art of Statistical Science: A Tribute to G. S. Watson
MASON, GUNST, and HESS · Statistical Design and Analysis of Experiments with Applications to Engineering and Science
McCULLOCH and SEARLE · Generalized, Linear, and Mixed Models
MURRAY · X-STAT 2.0 Statistical Experimentation, Design Data Analysis, and Nonlinear Optimization
PURI, VILAPLANA, and WERTZ · New Perspectives in Theoretical and Applied Statistics
RENCHER · Linear Models in Statistics
RENCHER · Methods of Multivariate Analysis
RENCHER · Multivariate Statistical Inference with Applications
ROSS · Introduction to Probability and Statistics for Engineers and Scientists

*Now available in a lower priced paperback edition in the Wiley Classics Library.

Texts, References, and Pocketbooks (Continued)

ROHATGI and SALEH · An Introduction to Probability and Statistics, *Second Edition*
RYAN · Modern Regression Methods
SCHOTT · Matrix Analysis for Statistics
SEARLE · Matrix Algebra Useful for Statistics
STYAN · The Collected Papers of T. W. Anderson: 1943–1985
TIAO, BISGAARD, HILL, PEÑA, and STIGLER (editors) · Box on Quality and Discovery: with Design, Control, and Robustness
TIERNEY · LISP-STAT: An Object-Oriented Environment for Statistical Computing and Dynamic Graphics
WONNACOTT and WONNACOTT · Econometrics, *Second Edition*
WU and HAMADA · Experiments: Planning, Analysis, and Parameter Design Optimization

JWS/SAS Co-Publications Section

KHATTREE and NAIK · Applied Multivariate Statistics with SAS Software, *Second Edition*
KHATTREE and NAIK · Applied Descriptive Multivariate Statistics Using SAS Software

WILEY SERIES IN PROBABILITY AND STATISTICS

ESTABLISHED BY WALTER A. SHEWHART AND SAMUEL S. WILKS

Editors
Robert M. Groves, Graham Kalton, J. N. K. Rao, Norbert Schwarz, Christopher Skinner

Survey Methodology Section

BIEMER, GROVES, LYBERG, MATHIOWETZ, and SUDMAN · Measurement Errors in Surveys
COCHRAN · Sampling Techniques, *Third Edition*
COUPER, BAKER, BETHLEHEM, CLARK, MARTIN, NICHOLLS, and O'REILLY (editors) · Computer Assisted Survey Information Collection
COX, BINDER, CHINNAPPA, CHRISTIANSON, COLLEDGE, and KOTT (editors) · Business Survey Methods
*DEMING · Sample Design in Business Research
DILLMAN · Mail and Telephone Surveys: The Total Design Method, *Second Edition*
DILLMAN · Mail and Internet Surveys: The Tailored Design Method
GROVES and COUPER · Nonresponse in Household Interview Surveys
GROVES · Survey Errors and Survey Costs
GROVES, BIEMER, LYBERG, MASSEY, NICHOLLS, and WAKSBERG · Telephone Survey Methodology
*HANSEN, HURWITZ, and MADOW · Sample Survey Methods and Theory, Volume I: Methods and Applications
*HANSEN, HURWITZ, and MADOW · Sample Survey Methods and Theory, Volume II: Theory
KISH · Statistical Design for Research

*Now available in a lower priced paperback edition in the Wiley Classics Library.

Survey Methodology (Continued)

*KISH · Survey Sampling

KORN and GRAUBARD · Analysis of Health Surveys

LESSLER and KALSBEEK · Nonsampling Error in Surveys

LEVY and LEMESHOW · Sampling of Populations: Methods and Applications, *Third Edition*

LYBERG, BIEMER, COLLINS, de LEEUW, DIPPO, SCHWARZ, TREWIN (editors) · Survey Measurement and Process Quality

SIRKEN, HERRMANN, SCHECHTER, SCHWARZ, TANUR, and TOURANGEAU (editors) · Cognition and Survey Research

VALLIANT, DORFMAN, and ROYALL · A Finite Population Sampling and Inference